RESEARCH AND PERSPECTIVES IN ALZHEIMER'S DISEASE
Fondation Ipsen

Editor

Yves Christen, Fondation Ipsen, Paris (France)

Editorial Board

Yves Agid, Hôpital Pitié Salpêtrière, Paris (France)
Albert Aguayo, McGill University, Montreal (Canada)
Luigi Amaducci, University of Florence, Florence (Italy)
Brian H. Anderton, Institute of Psychiatry, London (GB)
Raymond T. Bartus, Alkermes, Cambridge (USA)
Anders Björklund, University of Lund (Sweden)
Floyd Bloom, Scripps Clinic and Research Foundation, La Jolla (USA)
François Boller, Inserm U 324, Paris (France)
Carl Cotman, University of California, Irvine (USA)
Peter Davies, Albert Einstein College of Medicine, New York (USA)
André Delacourte, Inserm U 422, Lille (France)
Steven Ferris, New York University Medical Center, New York (USA)
Jean-François Foncin, Hôpital Pitié Salpêtrière, Paris (France)
Françoise Forette, Hôpital Broca, Paris (France)
Fred Gage, Salk Institute, La Jolla (USA)
Dmitry Goldgaber, State University of New York Stone Brook (USA)
John Hardy, Mayo Clinic, Jacksonville (USA)
Jean-Jacques Hauw, Hôpital Pitié Salpêtrière, Paris (France)
Claude Kordon, Inserm U 159, Paris (France)
Kenneth S. Kosik, Harvard Medical School, Center for Neurological
    Diseases and Brigham and Women's Hospital, Boston (USA)
Jacques Mallet, Hôpital Pitié Salpêtrière, Paris (France)
Colin L. Masters, University of Melbourne, Parkville (Australia)
Stanley I. Rapoport, National Institute on Aging, Bethesda (USA)
André Rascol, Hôpital Purpan, Toulouse (France)
Barry Reisberg, New York University Medical Center, New York (USA)
Allen Roses, Duke University Medical Center, Durham (USA)
Dennis J. Selkoe, Harvard Medical School, Center for Neurological
    Diseases and Brigham and Women's Hospital, Boston (USA)
Michael L. Shelanski, Columbia University, New York (USA)
Pierre-Marie Sinet, Hôpital Necker, Paris (France)
Peter St. George-Hyslop, University of Toronto, Toronto (Canada)
Robert Terry, University of California, La Jolla (USA)
Henry Wisniewski, Institute for Basic Research in Development Disabilities,
    Staten Island (USA)
Edouard Zarifian, Centre Hospitalier Universitaire, Caen (France)

Springer

*Berlin
Heidelberg
New York
Barcelona
Budapest
Hong Kong
London
Milan
Paris
Santa Clara
Singapore
Tokyo*

S.G. Younkin   R.E. Tanzi   Y. Christen (Eds.)

Presenilins and Alzheimer's Disease

With 17 Figures and 4 Tables

Springer

Younkin, S.G., M.D., Ph.D.
Mayo Clinic Jacksonville
4500 San Pablo Road
Jacksonville, FL 32224
USA

Tanzi, R.E., Ph.D.
Genetics and Aging Unit
and the Department of Neurology
Massachusetts General Hospital-East
Harvard Medical School
149, 13th Street
Boston, MA 02129-9142
USA

Christen, Y., Ph.D.
Fondation IPSEN
24, rue Erlanger
75651 Paris Cedex 16
France

ISSN 0945-6066
ISBN 3-540-63997-7   Springer-Verlag Berlin Heidelberg New York

Library of Congress Cataloging-in-Publication Data

Presenilins and Alzheimer's disease / S. G. Younkin, R. E. Tanzi, Y. Christen (eds.). p. cm. – (Research and perspectives in Alzheimer's disease) Includes bibliographical references and index. ISBN 3-540-63997-7 (hardcover) 1. Alzheimer's disease – Molecular aspects. 2. Alzheimer's disease – Genetic aspects. 3. Presenilins. I. Younkin, S. G. (Steven G.) II. Tanzi, Rudolph E. III. Christen, Yves. IV. Series. RC523.P74 1998 616.8'3107–dc21

This work is subject to copyright. All rights are reserved, whether the whole or part of the material is concerned, specifically the rights of translation, reprinting, reuse of illustrations, recitation, broadcasting, reproduction on microfilm or in any other way, and storage in data banks. Duplication of this publication or parts thereof is permitted only under the provisions of the German Copyright Law of September 9, 1965, in its current version, and permission for use must always be obtained from Springer-Verlag. Violations are liable for prosecution under the German Copyright Law.

© Springer-Verlag Berlin Heidelberg 1998
Printed in Germany

The use of general descriptive names, registered names, trademarks, etc., in this publication does not imply, even in the absence of a specific statement, that such names are exempt from the relevant protective laws and regulations and therefore free for general use.

Product Liability: The publishers cannot guarantee the accuracy of any information about dosage and application contained in this book. In every individual case the user must check such information by consulting the relevant literature.

Production: PRO EDIT GmbH, D-69126 Heidelberg

Cover design: Design & Production, D-69121 Heidelberg

Typesetting: Mitterweger Werksatz GmbH, Plankstadt

SPIN: 10551710   27/3136 – 5 4 3 2 1 0 – Printed on acid-free paper

Preface

Molecular and biochemical studies of Alzheimer's disease have recently undergone a major revolution with the discovery of the presenilin genes. Since 1995 when these genes were first identified to carry defects responsible for up to half of early onset familial Alzheimer's disease cases (Sherrington et al. 1995; Levy-Lahad et al. 1995), over 50 Alzheimer-associated mutations have been found in the presenilin genes, PS1 and PS2 (reviewed in Tanzi et al. 1996). Over 200 papers have been published regarding the characterization of the presenilins. Not since the amyloid β protein Precursor (APP) was isolated in 1987 (Kang et al. 1987; Goldgaber et al. 1987; Robakis et al. 1987; Tanzi et al. 1987) has the discovery of novel genes had such an impact on the field of Alzheimer's disease research. To whit, five separate sessions at the 1997 Society for Neuroscience Meeting are devoted solely to studies of the presenilins. The presenilins genes have clearly taken the field of Alzheimer's disease research by storm and appropriately so since defects in these genes can cause Alzheimer's disease as early as in one's late twenties.

One of the greatest revelations to arise from molecular studies of the presenilins is the finding that, like the familial Alzheimer's disease mutations in APP, the mutations in the presenilin genes lead to increased production and secretion of the longer form of the Aβ peptide, Aβ42 (Scheuner et al. 1995), which seeds β-amyloid formation in the brain. The presenilins have also been implicated in the process of programmed cell death (apoptosis; Kim et al. 1997; Wolozin et al. 1997). At the more basic biological level, these proteins have been linked to the developmental pathway involving the Notch genes (Levitan and Greenwald, 1995). These and other findings have provided great clues to the roles of the presenilin in both health and disease. It is thus appropriate that the Fondation IPSEN organized a meeting in July of 1997 to bring together a group of leaders in the area of presenilin biology to discuss ideas regarding the biological function of the presenilins and how defects in these genes cause the earliest onset form of Alzheimer's disease. An understanding of the mechanisms by which mutations in the presenilin genes cause neurodegeneration and dementia should greatly facilitate the development of novel strategies for treating Alzheimer's disease and related disorders.

January 1998

Rudolph TANZI
Steven YOUNKIN
Yves CHRISTEN

References

Goldgaber D, Lerman JI, McBride OW, Saffiotti U, Gajdusek DC (1987) Characterization of chromosomal localization of a cDNA encoding brain amyloid of fibril protein. *Science* 235:877–880

Kang J, Lemaire H, Unterbeck A, Salbaum JM, Masters CL, Grzeschik K, Multhaup G, Beyreuther K, Muller-Hill B (1987) The precursor of Alzheimer's disease amyloid A4 protein resembles a cell-surface receptor. *Nature* 325:733–736

Kim TW, Pettingell WH, Jung YK, Kovacs DM, Tanzi RE (1997) Alternative cleavage of Alzheimer-associated presenilins during apoptosis by a caspase-3 family protease. *Science* 277:373–376

Levitan D, Greenwald I (1995) Facilitation of lin-12-mediated signalling by sel-12, a *caenorhabditis elegans* S182 Alzheimer's disease gene. *Nature* 377:351–354

Levy-Lahad E, Wasco W, Poorkaj P, Romano DM, Oshima JM, Pettingell WH, Yu C, Jondro PD, Schmidt SD, Wang K, Crowley AC, Fu Y-H, Guenette SY, Galas D, Nemens E, Wijsman EM, Bird TD, Schellenberg GD, Tanzi RE (1995) Candidate gene for the chromosome 1 familial Alzheimer's disease locus. *Science* 269:973–977

Robakis NK, Ramakrishna N, Wolfe G, Wisniewski HM (1987) Molecular cloning and characterization of a cDNA encoding the cerebrovascular and the neuritic plaque amyloid peptides. *Proc Natl Acad Sci USA* 84:4190–4194

Scheuner D, Eckmann C, Jensen M, Song X, Citron M, Suzuki N, Bird TD, Hardy J, Hutton M, Kukull W, Larson E, Levy-Lahad E, Viitanen M, Peskind E, Poorkaj P, Schellenberg G, Tanzi RE, Wasco W, Lannfelt L, Selkoe D, Younkin S (1996) Secreted amyloid β-protein similar to that in the senile plaques of Alzheimer's disease is increased *in vivo* by the presenilin 1 and 2 and APP mutations linked to familial Alzheimer's disease. *Nature Med* 2:864–870

Sherrington R, Rogaev EI, Liang Y, Rogaeva EA, Levesque G, Ikeda M, Chi H, Lin C, Li G, Holman K, Tsuda T, Mar L, Foncin J-F, Bruni AC, Montesi MP, Sorbi S, Rainero I, Pinessi L, Nee L, Chumakov Y, Pollen D, Wasco W, Haines JL, Da Silva R, Pericak-Vance M, Tanzi RE, Roses AD, Fraser PE, Rommens JM, St George-Hyslop PH (1995) Cloning of a novel gene bearing missense mutations in early onset familial Alzheimer's disease. *Nature* 375:754–760

Tanzi RE, Gusella JF, Watkins PC, Bruns GAP, St George-Hyslop P, van Keuren ML, Patterson D, Pagan S, Kurnit DM, Neve RL (1987) Amyloid β protein gene: cDNA, mRNA distribution and genetic linkage near the Alzheimer locus. *Science* 235:880–884

Tanzi RE, Kovacs DM, Kim T-W, Moir RD, Guenette SY, Wasco W (1996) The gene defects responsible for familial Alzheimer's disease. *Neurobiol Dis* 3:159–168

Wolozin B, Iwasaki K, Vito P, Ganjei K, Lacana E, Sunderland T, Zhao B, Kusiak JW, Wasco W, D'Adamio L (1996) PS2 participates in cellular apoptosis: constitutive activity conferred by Alzheimer mutation. *Science* 274:1710–1713

Acknowledgments:

The editors wish to thank Mary Lynn Gage for editorial assistance and Jacqueline Mervaillie for the organization of the meeting in Paris.

Contents

Molecular Genetics of the Presenilins in Alzheimer's Disease
P.E. Fraser, G. Yu, G. Levesque, M. Ikeda, M. Nishimura, E. Rogaeva,
D. Westaway, P.H. St.George-Hyslop and G.A. Carlson 1

Alzheimer's Disease: A Matter of Dominance
J. Hardy and M. Hutton 11

Alternative Endoproteolysis of the Presenilins and Familial Alzheimer's
Disease
R.E. Tanzi, T.W. Kim, D.M. Kovacs, W. Wasco, R.D. Moir, W. Pettingell,
J. Henderson, and R. Mancini 19

The APP and PS1/2 Mutations Linked to Early Onset Familial Alzheimer's
Disease Increase the Extracellular Concentration of Aβ1-42(43) 27
S.G. Younkin

Metabolism and Function of Presenilin 1
S.S. Sisodia, G. Thinakaran, P.C. Wong, D.R. Borchelt, M.K. Lee, A. Doan,
J. Regard, H. Chen, H. Zheng, C. Eckman, H.H. Slunt, T. Ratovitsky,
F. Davenport, C. Harris, L.H.T. Van der Ploeg, S.G. Younkin, N.A. Jenkins,
N.G. Copeland, and D.L. Price 35

Mechanistic Studies of the Effect of Presenilins 1 and 2 on APP Metabolism
D.J. Selkoe, W. Xia, J. Zhang, M.B. Podlisny, C.A. Lemere, M. Citron,
and E.H. Koo 49

Presenilin 2 – APP Interactions
W. Wasco, R.E. Tanzi, R.D. Moir, A.C. Crowley, D.E. Merriam, D.M. Romano,
P.D. Jondro, and B.A. Kellerman 59

The Cellular Biology of Presenilin Proteins and a Novel Mechanism
of Amyloid β-Peptide Generation
C. Haass, J. Walter, A. Capell, C. Wild-Bode, J. Grünberg, T. Yamazaki,
I. Ihara, I. Zweckbronner, C. Jakubek, and R. Baumeister 71

Regulation of Presenilin 1 Phosphorylation and Transcriptional Activation
of Signal Transduction-Induced Genes by Muscarinic Receptors
U. Langer, C. Albrecht, M. Mayhaus, J. Velden, H. Wiegmann, J. Klaudiny,
D. Müller, H. von der Kammer, and R.M. Nitsch 79

Neuronal Regulation of Presenilin-1 Processing
H. Hartmann, J. Busciglio and B.A Yankner 85

Transgenic Approaches to the Study of Alzheimer's Disease
K. Duff 93

Subject Index 97

List of Contributors

Albrecht, C.
Center for Molecular Neurobiology, University of Hamburg, Martinistr. 52, 20246 Hamburg, Germany

Baumeister, R.
Laboratory of Molecular Biology/Genzentrum of the University of Munich, 81377 Munich, Germany

Borchelt, D.R.
Department of Pathology and the Neuropathology Laboratory, The John Hopkins University School of Medicine, Baltimore, MD 21205, USA

Busciglio, J.
Department of Neurology, Harvard Medical School and The Children's Hospital, Enders 260, 300 Longwood Avenue, Boston, MA 02115, USA

Capell, A.
Central Institute for Mental Health, Department of Molecular Biology, J5, 68159 Mannheim, Germany

Carlson, G.A.
McLaughlin Research Institute 1520, 23rd Street South, Great Falls, Montana 59409, USA

Chen, H.
The Neuropathology Laboratory, The John Hopkins University School of Medicine, Baltimore, MD 21205, USA

Citron, M.
Amgen, Inc., Thousand Oaks, CA 91320-1789, USA

Copeland, N.G.
Mammalian Genetics Laboratory, ABL-Basis Research Program, NCI-Frederick Cancer Center Research & Development, Frederick, MD 21701, USA

Crowley, A.C.
Genetics and Aging Unit and the Department of Neurology, Massachusetts General Hospital, Harvard Medical School, 149, 13th Street, Charlestown and Boston MA 02129-9142, USA

Davenport, F.
The Neuropathology Laboratory, The John Hopkins University School of Medicine, Baltimore, MD 21205, USA

Doan, A.
The Neuropathology Laboratory, The John Hopkins University School of Medicine, Baltimore, MD 21205, USA

Duff, K.
Mayo Clinic Jacksonville, 4500 San Pablo Rd., Jacksonville, FL 32224, USA

Eckman, C.
Mayo Clinic Jacksonville, 4500 San Pablo Rd., Jacksonville, FL 32224, USA

Fraser, P.E.
Centre for Research in Neurodegenerative Disease, Department of Medicine (Neurology), The Toronto Hospital, Tanz Neuroscience Bldg., 6, Queen's Park Crescent, Toronto, Ontario, Canada M5S 1A8

Grünberg, J.
Central Institute for Mental Health, Department of Molecular Biology, J5, 68159 Mannheim, Germany

Haass, C.
Central Institute for Mental Health, Department of Molecular Biology, J5, 68159 Mannheim, Germany

Hardy, J.
Mayo Clinic Jacksonville, 4500 San Pablo Rd., Jacksonville, FL 32224, USA

Harris, C.
The Neuropathology Laboratory, The John Hopkins University School of Medicine, Baltimore, MD 21205, USA

Hartmann, H.
Department of Neurology, Harvard Medical School and The Children's Hospital, Enders 260, 300 Longwood Avenue, Boston, MA 02115, USA

Henderson, J.
Genetics and Aging Unit and the Department of Neurology, Massachusetts General Hospital, Harvard Medical School, 149, 13th Street, Charlestown, MA 012129, USA

Hutton, M.
Mayo Clinic Jacksonville, 4500 San Pablo Rd., Jacksonville, FL 32224, USA

Ihara, I.
Department of Neuropathology, Institute for Brain Research, Faculty of Medicine, University of Tokyo, Tokyo 113, Japan

Ikeda, M.
Centre for Research in Neurodegenerative Disease, Department of Medicine (Neurology), The Toronto Hospital, Tanz Neuroscience Bldg., 6, Queen's Park Crescent, Toronto, Ontario, Canada M5S 1A8

Jakubek, C.
Laboratory of Molecular Biology/Genzentrum of the University of Munich, 81377 Munich, Germany

Jenkins, N.A.
Mammalian Genetics Laboratory, ABL-Basic Research Program, NCI-Frederick Cancer Center Research & Development, Frederick, MD 21701, USA

Jondro, P.D.
Genetics and Aging Unit and the Department of Neurology, Massachusetts General Hospital, Harvard Medical School, 149, 14th Street, Charlestown and Boston MA 02129-9142, USA

Kellermann, B.A.
Genetics and Aging Unit and the Department of Neurology, Massachusetts General Hospital, Harvard Medical School, 149, 13th Street, Charlestown and Boston MA 02129-9142, USA

Kim, T.W.
Genetics and Aging Unit and the Department of Neurology, Massachusetts General Hospital, Harvard Medical School, 149, 13th Street, Charlestown, MA 012129, USA

Klaudiny, J.
Center for Molecular Neurobiology, University of Hamburg, Martinistr. 52, 20246 Hamburg, Germany

Koo, E.H.
Department of Neuroscience, University of California, San Diego, CA 92093-0691, USA

Kovacs, D.M.
Genetics and Aging Unit and the Department of Neurology, Massachusetts General Hospital, Harvard Medical School, 149, 13th Street, Charlestown, MA 012129, USA

Langer, U.
Center for Molecular Neurobiology, University of Hamburg, Martinistr. 52, 20246 Hamburg, Germany

Lee, M.K.
Department of Pathology and the Neuropathology Laboratory, The John Hopkins University School of Medicine, Baltimore, MD 21205, USA

Lemere, C.A.
Center for Neurologic Diseases, Brigham and Women's Hospital, Department of Neurology, Harvard Medical School, 77 Avenue Louis Pasteur, Boston, MA 02115, USA

Levesque, G.
Centre for Research in Neurodegenerative Disease, Department of Medicine (Neurology), The Toronto Hospital, Tanz Neuroscience Bldg., 6, Queen's Park Crescent, Toronto, Ontario, Canada M5S 1A8

Mancini, R.
Genetics and Aging Unit and the Department of Neurology, Massachusetts General Hospital, Harvard Medical School, 149, 13th Street, Charlestown, MA 012129, USA

Mayhaus, M.
Center for Molecular Neurobiology, University of Hamburg, Martinistr. 52, 20246 Hamburg, Germany

Merriam, D.E.
Genetics and Aging Unit and the Department of Neurology, Massachusetts General Hospital, Harvard Medical School, 149, 13th Street, Charlestown and Boston MA 02129-9142, USA

Moir, R.D.
Genetics and Aging Unit and the Department of Neurology, Massachusetts General Hospital, Harvard Medical School, 149, 13th Street, Charlestown and Boston MA 02129-9142, USA

Müller, D.
Center for Molecular Neurobiology, University of Hamburg, Martinistr. 52, 20246 Hamburg, Germany

Nishimura, M.
Centre for Research in Neurodegenerative Disease, Department of Medicine (Neurology), The Toronto Hospital, Tanz Neuroscience Bldg., 6, Queen's Park Crescent, Toronto, Ontario, Canada M5S 1A8

Nitsch, R.M.
Center for Molecular Neurobiology, University of Hamburg, Martinistr. 52, 20246 Hamburg, Germany

Pettingell, W.
Genetics and Aging Unit and the Department of Neurology, Massachusetts General Hospital, Harvard Medical School, 149, 13th Street, Charlestown, MA 012129, USA

Podlisny, M.B.
Center for Neurologic Diseases, Brigham and Women's Hospital, Department of Neurology, Harvard Medical School, 77 Avenue Louis Pasteur, Boston, MA 02115, USA

Price, D.L.
Departments of Pathology, Neurology and Neuroscience, and The Neuropathology Laboratory, The John Hopkins University School of Medicine, Baltimore, MD 21205, USA

Ratovitsky, T.
The Neuropathology Laboratory, The John Hopkins University School of Medicine, Baltimore, MD 21205, USA

Regard, J.
Department of Neuroscience, The John Hopkins University School of Medicine, Baltimore, MD 21205, USA

Rogaeva, E.
Centre for Research in Neurodegenerative Disease, Department of Medicine (Neurology), The Toronto Hospital, Tanz Neuroscience Bldg., 6, Queen's Park Crescent, Toronto, Ontario, Canada M5S 1A8

Romano, D.M.
Genetics and Aging Unit and the Department of Neurology, Massachusetts General Hospital, Harvard Medical School, 149, 13th Street, Charlestown and Boston MA 02129-9142, USA

Selkoe, D.J.
Center for Neurologic Diseases, Brigham and Women's Hospital, Department of Neurology, Harvard Medical School, 77 Avenue Louis Pasteur, Boston, MA 02115, USA

Sisodia, S.S.
Departments of Pathology, Neuroscience and the Neuropathology Laboratory, The John Hopkins University School of Medicine, Baltimore, MD 21205, USA

Slunt, H.H.
The Neuropathology Laboratory, The John Hopkins University School of Medicine, Baltimore, MD 21205, USA

St.George-Hyslop, P.H.
Centre for Research in Neurodegenerative Disease, Department of Medicine (Neurology), The Toronto Hospital, Tanz Neuroscience Bldg., 6, Queen's Park Crescent, Toronto, Ontario, Canada M5S 1A8

Tanzi, R.E.
Genetics and Aging Unit and the Department of Neurology, Massachusetts General Hospital, Harvard Medical School, 149, 13th Street, Charlestown and Boston MA 02129-9142, USA

Thinakaran, G.
Department of Pathology and the Neuropathology Laboratory, The John Hopkins University School of Medicine, Baltimore, MD 21205, USA

Van der Ploeg, L.H.T.
Merck Research Laboratories, Rahway, New Jersey 07065, USA

Velden, J.
Center for Molecular Neurobiology, University of Hamburg, Martinistr. 52, 20246 Hamburg, Germany

von der Kammer, H.
Center for Molecular Neurobiology, University of Hamburg, Martinistr. 52, 20246 Hamburg, Germany

Walter, J.
Central Institute for Mental Health, Department of Molecular Biology, J5, 68159 Mannheim, Germany

Wasco, W.
Genetics and Aging Unit and the Department of Neurology, Massachusetts General Hospital, Harvard Medical School, 149, 13th Street, Charlestown and Boston MA 02129-9142, USA

Westaway, D.
Centre for Research in Neurodegenerative Disease, Department of Medicine (Neurology), The Toronto Hospital, Tanz Neuroscience Bldg., 6, Queen's Park Crescent, Toronto, Ontario, Canada M5S 1A8

Wiegmann, H.
Center for Molecular Neurobiology, University of Hamburg, Martinistr. 52, 20246 Hamburg, Germany

Wild-Bode, C.
Central Institute for Mental Health, Department of Molecular Biology, J5, 68159 Mannheim, Germany

Wong, P.C.
Department of Pathology and the Neuropathology Laboratory, The John Hopkins University School of Medicine, Baltimore, MD 21205, USA

Xia, W.
Center for Neurologic Diseases, Brigham and Women's Hospital, Department of Neurology, Harvard Medical School, 77 Avenue Louis Pasteur, Boston, MA 02115, USA

Yamazaki, T.
Department of Neuropathology, Institute for Brain Research, Faculty of Medicine, University of Tokyo, Tokyo 113, Japan

Yankner, B.A.
Department of Neurology, Harvard Medical School and The Children's Hospital, Enders 260, 300 Longwood Avenue, Boston, MA 02115, USA

Younkin, S.G.
Mayo Clinic Jacksonville, 4500 San Pablo Rd., Jacksonville, FL 32224, USA

Yu, G.
Centre for Research in Neurodegenerative Disease, Department of Medicine (Neurology), The Toronto Hospital, Tanz Neuroscience Bldg., 6, Queen's Park Crescent, Toronto, Ontario, Canada M5S 1A8

Zhang, J.
Center for Neurologic Diseases, Brigham and Women's Hospital, Department of Neurology, Harvard Medical School, 77 Avenue Louis Pasteur, Boston, MA 02115, USA

Zheng, H.
Merck Research Laboratories, Rahway, New Jersey 07065, USA

Zweckbronner, I.
Laboratory of Molecular Biology/Genzentrum of the University of Munich, 81377 Munich, Germany

Molecular Genetics of the Presenilins in Alzheimer's Disease

P. E. Fraser, G. Yu, G. Levesque, M. Ikeda, M. Nishimura, E. Rogaeva,
D. Westaway, P. H. St. George-Hyslop and G. A. Carlson[*]

Molecular Genetics of Alzheimer's Disease

Molecular genetic studies in pedigrees with a reasonably unambiguous single gene autosomal dominant pattern of inheritance have led to the identification of the four different genes associated with inherited susceptibility to Alzheimer's disease (AD) – namely β-amyloid precursor protein (βAPP; Goate et al. 1991; Murrell et al. 1991; Naruse et al. 1991; Hendricks et al. 1992; Karlinsky et al. 1992; Mullan et al. 1992), apolipoprotein E (APOE; Saunders et al. 1993), presenilin 1 (PS1; Sherrington et al. 1995), and presenilin 2 (PS2; Rogaev et al. 1995). There is also genetic evidence that some of these genes may function within the same biochemical pathway and have additive effects. Thus, carriers of $\beta APP_{Val717Ile}$ who have one or more ε4 alleles at APOE have an earlier onset than relatives with the same βAPP mutation and the ε2 allele at APOE (St. George-Hyslop et al. 1994; Nacmias et al. 1995; Sorbi et al. 1995). However, because a significant number of pedigrees multiply affected by AD have been found which do not segregate mutations/polymorphisms in any of the four known AD susceptibility genes, it is suspected that at least one and possible several additional AD susceptibility genes must also exist but have not yet been identified (Sherrington et al. 1996).

Presenilin 1

After the discovery that βAPP missense mutations were quite rare as a cause of AD (Tanzi et al. 1992), several groups undertook a survey of the remaining non-sex-linked chromosomes other than chromosomes 19 and 21. These studies identified a series of polymorphic genetic markers located on chromosome 14q24.3 (D14S43, D14S71, D14S77 and D14S53) which showed robust evidence of linkage to an early onset from of familial AD, $z \geq 23.0$ at $\theta = 0.01$ (Schellenberg et al. 1992; St. George-Hyslop et al. 1992; Van Broeckhoven et al. 1992). Subse-

---

[*] Centre for Research in Neurodegenerative Disease, Department of Medicine, University of Toronto, Department of Medicine (Neurology), The Toronto Hospital, Tanz Neuroscience Bldg., 6, Queen's Park Crescent, Toronto, Ontario, CANADA M5S 1A8
McLaughlin Research Institute, Great Falls, Montana

quent genetic mapping studies narrowed the region containing this third Alzheimer susceptibility locus (AD3) to a region of approximately 10 centiMorgans between the marker D14S271 at the centromeric end and D14S53 at the telomeric end, a physical distance of approximately 7 megabases. The actual disease gene (presenilin 1; PS1) was isolated using a positional cloning strategy (Sherrington et al. 1995) and is a novel gene that is highly conserved in evolution, being present in *C. elegans* (Levitan and Greenwald 1995) and *D. melanogaster* (Boulianne et al. 1997), and appears to encode a polytopic integral membrane protein.

To date, more than 35 different mutations have been discovered in the PS1 gene (Table 1). The majority of these mutations are missense mutations giving rise to the substitution of a single amino acid. These mutations are predominantly located in highly conserved transmembrane domains, at or near putative membrane interfaces, or in the N-terminal hydrophobic or C-terminal hydrophobic residues of the putative TM6-TM7 loop domain. A single splicing defect mutation has been identified which involves a point mutation in the splice acceptor site at the 5' end of exon 10 (Perez-Tur et al. 1996; Sato et al. 1997; Kwok et al., 1997). No deletions, rearrangements or nonsense mutations resulting in truncated proteins have yet been found in AD-affected subjects. The absence of the latter types of mutations, all of which would cause loss of function effects, raises the question as to whether such mutations might be lethal or might lead to other disease genotypes. However, it is of note that mice with heterozygous knock out of the murine PS1 gene are phenotypically normal (Wong et al. 1996).

Presenilin 2

During the cloning of the PS1 gene on chromosome 14, a very similar sequence was identified in the public nucleotide sequence databases (Rogaev et al. 1995). Further analysis revealed that this similar nucleotide sequence was derived from a gene on chromosome 1 and encodes a polypeptide whose open reading frame contained 448 amino acids with substantial amino acid sequence identity with that of the PS1 protein (overall identity approximately 60%), and many of the intron exon boundaries (especially those relating to the highly conserved transmembrane (TM) domains) are identical between this gene and PS1 (Levy-Lahad et al. 1996; Sherrington et al. 1996). Cumulatively, these observations therefore suggested that this novel gene (presenilin 2; PS2) was a homologue of PS1 gene on chromosome 14.

Mutational analyses uncovered two different missense mutations in the PS2 gene in families segregating early-onset forms of AD (Table 1). The first mutation (Asn141Ile) was detected in a proportion of families of Volga German ancestry (Levy-Lahad et al. 1995; Rogaev et al. 1995), in which the familial AD (FAD) locus had been independently mapped by genetic linkage studies to chromosome 1 (Levy-Lahad et al. 1995). The second mutation (Met239Val) was discovered in an Italian pedigree (Rogaev et al. 1995) and affects a residue (Met239) that is also mutated in PS1 (Met233; Kwok et al. 1997). However, in contrast to the frequency

Table 1. Missense mutations in the presenilin genes. Mutations in the PS1 and PS2 genes. The majority of the mutations are located in or near putative TM domains. Certain models predict that many of the mutations would be aligned on the internal face of intramembrane helix bundles except were noted by 1. Mutations marked by 2 have been shown to produce high levels of $A\beta_{1-42}$ in plasma and from fibroblasts. The PS1 residues equivalent to PS2 mutations are denoted 3 = equivalent to Asn_{135} in PS1; 4 = equivalent to Met_{233} in PS1. The *sel-12* Cys60Ser loss of function mutation affects the equivalent Cys_{92} in PS1. Mutation 6 creates putative glycosylation site; 7 causes a PS1 Exon 10 splicing defect that inhibits physiologic endoproteolytic cleavage

		Presenilin 1 (S182)	
Codon	Location	Mutation	Phenotype
79	N-term loop	Ala→?	FAD, onset 64 years (personal comm)
82[1]	TM1	Val→Leu	FAD, onset 55 years (Campion et al. 1995)
96	TM1	Val→Phe	FAD (Kamino et al. 1996)
115	TM1→TM2 loop	Tyr→His	FAD, onset 37 years (Campion et al. 1995)
120	TM1→TM2 loop	Glu→Asp	FAD, onset 48 years
139	TM2	Met→Thr	FAD, onset 49 years (Campion et al. 1995)
139	TM2	Met→Val	FAD, onset 40 years (The Alz. Dis. Collab. Group 1995)
143	TM2	Ile→Thr	FAD, onset 35 years (Cruts et al. 1995)
146	TM2	Met→Leu	FAD, onset 45 years (Sherrington et al. 1995)
146[2]	TM2	Met→Val	FAD, onset 38 years (The Alz. Dis. Collab. Group 1995)
146	TM2	Met→Ile	FAD, onset 40 years
163	TM3 interface	His→Arg	FAD, onset 50 years (Sherrington et al. 1995)
163	TM3 interface	His→Tyr	FAD, onset 47 years (The Alz. Dis. Collab. Group 1995)
171	TM3	Leu→Pro	FAD, onset 40 years (Ramirez-Duenas et al., 1997)
209[2]	TM4 interface	Gly→Val	FAD (Kamino et al. 1996)
213	TM4 interface	Ile→Thr	FAD (Kamino et al. 1996)
231[1]	TM5	Ala→Thr	FAD, onset 52 years (Campion et al. 1995)
233[4]	TM5	Met→Thr	FAD, onset 35 years (Kwok et al., 1997)
235	TM5	Leu→Pro	FAD, onset 32 years (Campion et al. 1996)
246	TM6	Ala→Glu	FAD, onset 55 years (Sherrington et al. 1995)
260	TM6	Ala→Val	FAD, onset 40 years (Rogaev et al. 1995)
263	TM6→TM7 loop	Cys→Arg	FAD, onset 47 years
264	TM6→TM7 loop	Pro→Leu	FAD, onset 45 years (Campion et al. 1995)
267	TM6→TM7 loop	Pro→Ser	FAD, onset 35 years (The Alz. Dis. Collab. Group 1995)
280[6]	TM6→TM7 loop	Glu→Ala	FAD, onset 47 years (The Alz. Dis. Collab. Group 1995)
280	TM6→TM7 loop	Glu→Gly	FAD, onset 42 years (The Alz. Dis. Collab. Group 1995)
285	TM6→TM7 loop	Ala→Val	FAD, onset 50 years (Rogaev et al. 1995)
286	TM6→TM7 loop	Leu→Val	FAD, onset 50 years (Sherrington et al. 1995)
del291-319[7]	TM6→TM7 loop	short loop	FAD (Perez-Tur et al. 1996; Sato et al. 1996; Kwok et al., 1997)
384	TM6→TM7 loop	Gly→Ala	FAD, onset 35 years (Cruts et al. 1995)
392	TM6→TM7 loop	Leu→Val	FAD, onset 25-40 years (Rogaev et al. 1995)
410	TM7	Cys→Tyr	FAD, onset 48 years (Sherrington et al. 1995)

Table 1. continue

Codon	Location	Presenilin 2 Mutation	Phenotype
141[3]	TM2	Asn→Ile	FAD, onset 50–65 years (Levy-Lahad et al. 1995; Rogaev et al. 1995)
239[4]	TM5	Met→Val	FAD, onset variable 45–85 years (Rogaev et al. 1995)

of PS1 mutations, screening of large data sets reveals that PS2 mutations are likely to be rare (Sherrington et al. 1996).

Other Genes

Several large surveys of subjects with FAD have indicated that the four currently known AD susceptibility loci do not account for the disease in all pedigrees (Sherrington et al. 1996). Since pedigrees lacking mutations in any of the four known AD genes do not have a singular phenotype, but instead comprise a mix of early-onset autosomal dominant pedigrees and late-onset multiplex pedigrees, it is likely that there are several FAD genes remaining to be found. Some of these FAD loci will probably be associated with rather rare but highly penetrant defects, similar to those seen with mutations in PS1 and βAPP. Other genes may result in incompletely penetrant autosomal dominant traits like that associated with PS2, while still others may be AD susceptibility genes similar to APOE.

Several association studies have been performed on likely candidate genes, including α1-anti-chymotrypsin (Kamboh et al. 1995) and the VLDL receptor (Okuizumi et al. 1995). However, follow-up studies have not generated consistent confirmation of these associations (e.g., no replication was found of the α1-anti-chymotrypsin association in a large follow-up study; Haines et al. 1996), and so other loci are currently being sought. Recently, genetic linkage studies in a large data set of pedigrees with late onset AD have provided provisional evidence for the existence of another AD susceptibility locus on chromosome 12.

Biology of the Human Presenilins

In an attempt to elucidate the biological functions of the presenilin proteins, we and others have generated a series of antibodies directed at different epitopes of PS1 and PS2. The most convincing of these antibodies have been those directed at the N-terminus (such as the antibody AB14; Thinakaran et al. 1996b) and to the large hydrophilic loop domain between residues 260 and 409 of PS1 (or the equivalent sequences in PS2; Walter et al. 1996). Antibodies to the highly conserved C-terminus, while clearly recognizing the C-terminus of both PS1 and PS2, generally also depict several other bands on Western blots, the specificity of which is in doubt. The N-terminal and loop antibodies have been used by our group to show that the presenilin proteins are predominantly located in the perinuclear envelope and the contiguous endoplasmic reticulum, the Golgi apparatus, and some as yet uncharacterized cytoplasmic vesicles (Walter et al. 1996; De Strooper et al. 1997; Fig. 1). This distribution of PS1 and PS2 immunoreactivity has been observed in a number of different transfected cell types as well as in native cell types such as fibroblasts or explanted hippocampal neurons from embryonic mice (De Strooper et al., in preparation). The association of presenilin immunoreactivity with intracellular membranes has also been confirmed by electron microscopy (Fraser et al., unpublished). In addition to the predominantly perinuclear distribution of presenilin immunoreactivity that is observed in non-neuronal cells, we have also recently shown that there is an aggregation of presenilin immunoreactivity in growth cones of cultured hippocampal neurons (De Strooper et al., in preparation). It is important to note that the distribution of presenilin immunoreactivity is essentially identical regardless of whether the antibodies are directed to the N-terminus or to residues after residue 290. This latter observation is important because we and others have shown that both PS1 and PS2 undergo an endoproteolytic cleavage near residue 290 (Thinakaran et al. 1996a, b). Furthermore, under basal conditions in non-transfected cells, the

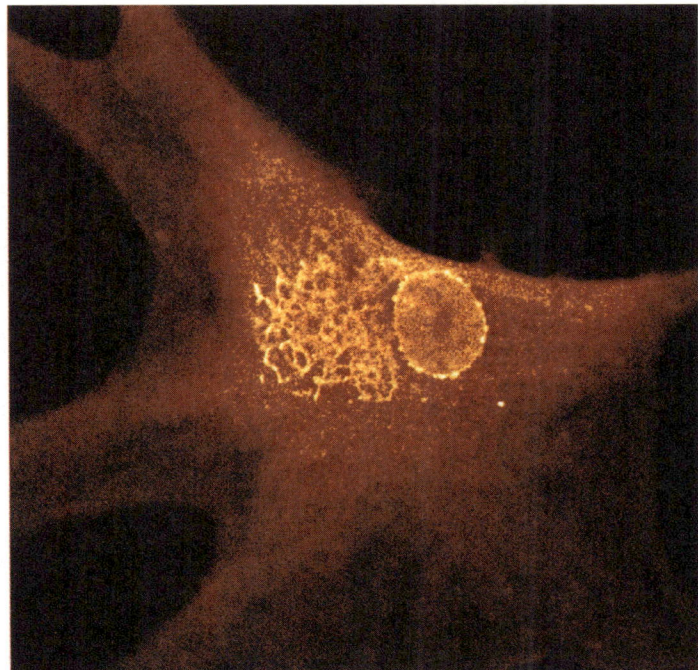

Fig. 1. Laser confocal micrograph (64 X magnification) of a human native fibroblast stained with the N-terminal PS1 antibody Ab14. PS1 immunoreactivity is predominantly located in the ER, perinuclear envelope, Golgi and some cytoplasmic vesicles. Little or no immunoreactivity is seen at the cell surface

majority of PS1 immunoreactivity detectable on Western blots consists of a N-terminal 35 kDa and a C-terminal 18 kDa fragments. Thus, the predominant immunoreactivity detected by these antibodies in cultured cells probably reflects mostly the distribution of the N- and C-terminal cleavage products. It is unclear whether the distribution of the endoproteolytic cleavage products mirrors, or differs from, the distribution of the precursor holoprotein.

To explore the topology of the presenilin proteins, we have also undertaken studies using immunocytochemistry and differential solubilization of cellular membranes with digitonin and saponin (De Strooper et al. 1997). These studies reveal that the N-terminus and the large hydrophilic loop domain between residues 260 and 409 of PS1 are oriented into the cytoplasm. Conversely, the second hydrophilic domain between TM1 and TM2 is located within the luminal side of intracellular membrane structures. In our hands, the orientation of the C-terminus has been ambiguous. In conjunction with hydrophobicity analyses, these studies suggest that there are minimally six TM domains between the N-terminus and the exposed large hydrophilic loop domain from residue 260 to 409. These studies do not, however, resolve whether there are additional TM domains toward the C-terminus of the presenilins. Consequently, current models suggesting six, seven, eight or nine TM domains would be compatible with our experimental observations. Others, however, have suggested that there are eight TM domains (Thinakaran et al. 1996a, b).

The observations that the exposed loop domain between residues 260 and 409 is located in the cytoplasm, is the site of endoproteolytic cleavage, is the site of alternate splicing (Rogaev et al. 1995, 1997), and is the site of several pathogenic mutations including a splicing mutation have cumulatively led to the speculation that this loop domain may be a functional region of the presenilin proteins. We have therefore begun to use this loop domain in yeast-two-hybrid interactionassays to identify proteins that have a functional interaction with the presenilins. Such studies may be helpful in identifying the biochemical pathways in which the presenilins function. While these studies are still incomplete, the two initial interacting proteins, depicted by clone GT24 (Levesque et al., unpublished data) and by clones PS1LY2H-29 and -31 (Fraser et al., unpublished data), seem to indicate, respectively, that the presenilins may function both in intracellular signal transduction during embryogenesis and may have a role in protein processing.

Animal Models Using Presenilin Sequences

To attempt to generate an animal model that replicates the human illness, we are creating animal models that either overexpress a human mutant PS1 or PS2 transgenes or express an endogenous murine presenilin gene that has been modified by homologous recombination so that it contains a missense mutation corresponding to a human FAD-related mutation. To date, careful sequential analysis of behaviour, electrophysiology, neuropathology, and neurochemistry of the brain of transgenic mice with mutant PS1 transgenes up to one year of age has revealed two subtle differences in comparison to transgenic mice overexpressing wild-type human PS1 and non-transgenic littermates. First, there is a modest increase in the ratio of $A\beta_{42}/A\beta_{total}$ in the brain of transgenic mice with mutant human PS1 transgenes (Citron et al. 1997) which is not related to the transgene copy number or insertion site and which is similar to that observed both in transfected cells expressing mutant PS1 cDNAs and in cultured fibroblasts and plasma from humans with PS1 mutations (Martin et al. 1995; Scheuner et al. 1996; Citron et al. 1997). Despite this alteration in βAPP processing, which is compatible with a postulated role for the presenilins in protein trafficking, by one year of age there has been no evidence of overt deposition of Aβ peptide in the brains of the transgenic mice with mutant human PS1 transgenes and wild type endogenous murine βAPP or wild type human βAPP transgenes. Secondly, we have observed electrophysiological evidence of disturbed neuronal function as manifested by an enhancement of long-term depression (LTD) and a reduction in long-term potentiation (LTP) in hippocampal slices from several different lines of transgenic mice bearing mutant human PS1 sequences (Agopyian et al., unpublished data). These electrophysiological changes were not observed in control mice overexpressing a wild-type human PS1 transgene or in non-transgenic littermates. These changes in electrophysiological correlates of memory and learning are accompanied by a subtle behavioral difference in the mutant PS1 transgenic mice manifested by longer swim paths and faster swim speeds in the Morris water maze test using a hidden platform. These differences were not apparent when the platform was visible and could not be attributed to any locomotor, sensory or musculo-skeletal differences from control mice. It remains to be determined whether these changes in electrophysiology and behaviour are progressive, and whether they related to the modest increase in $A\beta_{42}/A\beta_{total}$ ratios or whether they reflect another intrinsic property of overexpression of a mutant PS1 transgene. Studies of mice created with targeted mutagenesis by homologous recombination and studies of mice with mutant PS1 sequences but null mutations of the endogenous murine βAPP genes are underway to address these questions.

Acknowledgments. Supported by grants from the Medical Research Council of Canada, the Canadian Genetic Diseases Network, the Alzheimer Association of Ontario and its Member Chapters, the American Health Assistance Foundation, the EJLB Foundation, the Scottish Rite Charitable Foundation, the Ontario Mental Health Foundation, and by Fellowship support from the Peterborough-Burgess Fund, the Helen B. Hunter Fellowship, and the Howard Hughes Medical Research Foundation.

References

Boulianne G, Livne-Bar I, Humphreys JM, Rogaev E, St George-Hyslop P (1997) Cloning and mapping of a close homologue of human presenilins in D. melanogaster. NeuroReport 8: 1025–1029

Campion D, Brice A, Dumanchin C, Puel M, Baulac M, Clerget-Darpoux F, Agid Y, Frebourg T (1996) A novel presenilin 1 mutation in familial Alzheimer's disease with onset age of 29 years. NeuroReport 7: 1582–1584

Campion D, Flaman J-M, Brice A, Hannequin D, Dubois B, Martin C, Moreau V, Charbonnier F, Didierjean O, Tardieu S, Mallet J, Bellis M, Clerget-Darpoux C, Agid Y, Frebourg T (1995) Mutations of the Presenilin-1 gene in families with early onset Alzheimer's disease. Hum Molec Genet 4: 2373–2377

Citron M, Westaway D, Xia W, Carlson G, Diehl TS, Levesque G, Johnson-Wood K, Lee M, Seubert P, Davis A, Kholodenko D, Motter R, Sherrington R, Perry B, Yao H, Strome R, Lieberburg I, Rommens J, Kim S, Schenk D, Fraser P, St George-Hyslop P, Selkoe DJ (1997) Mutant presenilins of Alzheimer's disease increase production of 42 residue amyloid β-protein in both transfected cells and transgenic mice. Nature Med 3: 67–72

Cruts M, Martin J-J, Van Broeckhoven C (1995) Molecular genetic analysis of familial early-onset Alzheimer's disease linked to chromosome 14q24.3. Hum Molec Genet 4: 2363–2371

De Strooper B, Beullens M, Contreras B, Craessaerts K, Moechars D, Bollen M, Fraser P, St George-Hyslop P, Van Leuven F (1997) Postranslational modification, subcellular localization and membrane orientation of the Alzheimer's disease associated presenilins. J Biol Chem 272: 3590–3598

Goate AM, Chartier-Harlin M-C, Mullan M, Brown J, Crawford F, Fidani L, Guiffra L, Haynes A, Irving N, James L, Mant R, Neuton P, Rooke K, Roques P, Talbot C, Pericak-Vance M, Roses A, Williamson R, Rossor M, Owen M, Hardy JA (1991) Segregation of a missense mutation in the amyloid precursor protein gene with familial Alzheimer disease. Nature 349: 704–706

Haines JL, Pritchard ML, Saunders AM, Schildkraut JM, Growdon JH, Gaskell P, Farrer LA, Auerbach SA, Gusella JF, Yamoaka L, Conneally PM, Roses AD, Pericak-Vance MA (1996) No genetic effect of alpha-1 antichymotrypsin in Alzheimer disease. Genomics 33: 53–56

Hendricks M, van Duijn CM, Cras P, Cruts M, van Hul W, Van Harskamp F, Warren A, McInnis M, Antonarakis G, Martin J-J, Hofman A, Van Broeckhoven C (1992) Presenile dementia and cerebral hemorrhage linked to a mutation at codon 692 of the β-amyloid precursor protein gene. Nature Gene 1: 218–221

Kamboh MI, Sanghera DK, Ferrell RE, DeKosky ST (1995) ApoE ϵ4 associated Alzheimer's disease risk is modified by α_1 anti-chymotrypsin polymorphism. Nature Genet 10: 486–488

Kamino K, Sato S, Sakaki Y, Yoshiiwa A, Nishiwaki Y, Takeda M, Tanabe H, Nishimura T, Li K, St George-Hyslop P, Miki T, Ohihara T (1996) Three different mutations of the presenilin 1 gene in early onset Alzheimer's disease families. Neurosci Lett 208: 195–198

Karlinsky H, Vaula G, Haines JL, Ridgley J, Bergeron C, Mortilla M, Tupler R, Percy M, Robitaille Y, Crapper MacLachlan DR, St George-Hyslop P (1992) Molecular and prospective phenotypic characterization of a pedigree with familial Alzheimer disease and a missense mutation in codon 717 of the β-amyloid precursor protein (APP) gene. Neurology 42: 1445–1453

Kwok JBJ, Taddel K, Hallupp M, Fisher C, Brook WS, Broe G, Hardy J, Fulham MJ, Nicholson GA, Stell R, St George Hyslop PH, Fraser PE, Kakulas B, Clarnette R, Relkin N, Gandy SE, Schofield PR, Martins RN (1997) Two Novel (M233T and R278T) presenilin 1 mutations in early onset Alzheimer's disease and preliminary evidence for association of presenilin1 mutations with a novel phenotype. NeuroReport 8: 1537–1542

Levitan D, Greenwald I (1995) Facilitation of lin-12-mediated signalling by sel-12, a Caenorhabditis elegans S182 Alzheimer's disease gene. Nature 377: 351–354.

Levy-Lahad E, Wijsman EM, Nemens E, Anderson L, Goddard KAB, Weber JL, Bird TD, Schellenberg GD (1995) A familial Alzheimer's disease locus on chromosome 1. Science 269: 970–973

Levy-Lahad E, Poorkaj P, Wang K, Fu YH, Oshima J, Mulligan J, Schellenberg GD (1996) Genomic structure and expression of STM2, the chromosome 1 familial Alzheimer disease gene. Genomics 34: 198–204

Martin RN, Turner BA, Carroll RT, Sweeney D, Kim KS, Wisniewski HM, Blass JP, Gibson GE, Gandy SE (1995) High levels of amyloid β-protein from S182 (Glu246) familial Alzheimer's cells. NeuroReport 7: 217–220

Mullan MJ, Crawford F, Axelman K, Houlden H, Lilius L, Winblad B, Lannfelt L, Hardy J (1992) A pathogenic mutation for probable Alzheimer's disease in the APP gene at the N-terminus of β-amyloid. Nature Genet 1: 345-347

Murrell J, Farlow M, Ghetti B, Benson MD (1991) A mutation in the amyloid precursor protein associated with hereditary Alzheimer's disease. Science 254: 97-99

Nacmias B, Latteraga S, Tulen P, Piacentini S, Bracco L, Amaducci L, Sorbi S (1995) ApoE genotype and familial Alzheimer's disease: a possible influence on age-of-onset in APP71Val→Ile mutated families. Neurosci Lett 183: 1-3

Naruse S, Igarashi S, Kobyashi H, Aoki K, Inuzuki I, Kaneko K, Shimizu T, Iihara K, Kojima T, Miyatake T, Tsuji S (1991) Missense mutation (Val→Ile) in exon 17 of the amyloid precursor protein gene in Japanese familial Alzheimer disease. Lancet 337: 978-979

Okuizumi K, Onodera O, Namba Y, Ikeda K, Yamamoto T, Seki K, Ueki A, Nanko S, Tanaka H, Takahashi H, Tsuji S (1995) Genetic association of the very low density lipoprotein (VLDL) receptor gene with sporadic AD. Nature Genet 11: 207-209

Perez-Tur J, Froelich S, Prihar G, Crook R, Baker M, Duff K, Wragg M, Hardy J, Goate A, Lannfelt L, Hutton M (1996) A mutation in Alzheimer's disease destroying a splice acceptor site in the presenilin 1 gene. NeuroReport 7: 297-301

Rogaev EI, Sherrington R, Rogaeva EA, Levesque G, Ikeda M, Liang Y, Chi H, Lin C, Holman K, Tsuda T, Mar L, Sorbi S, Nacmias B, Piacentini S, Amaducci L, Chumakov I, Cohen D, Lannfelt L, Fraser PE, Rommens JM, St. George-Hyslop P (1995) Familial Alzheimer's disease in kindreds with missense mutations in a novel gene on chromosome 1 related to the Alzheimer's disease type 3 gene. Nature 376: 775-778

Rogaev EI, Sherrington R, Wu C, Levesque G, Liang Y, Rogaeva EA, Chi H, Ikeda M, Holman K, Lin C, Lukiw WJ, de Jong PJ, Fraser PE, Rommens JM, St George-Hyslop PH (1997) Analysis of the 5' sequence, genomic structure and alternative splicing of the presenilin 1 gene associated with early onset Alzheimer's disease. Genomics, 40: 415-424

Sato S, Kamino K, Miki T, Doi A, Li K, St. George-Hyslop P, Ogihara T and Sakaki Y (1997) Splicing mutation of presenilin 1 gene for early onset familial Alzheimer's disease. Hum Mutation, in press

Saunders A, Strittmatter WJ, Schmechel S, St George-Hyslop P, Pericak-Vance M, Joo SH, Rosi BL, Gusella JF, Crapper-McLachlan D, Growden J, Alberts MJ, Hulette C, Crain B, Goldgaber D, Roses AD (1993) Association of apolipoprotein E allele ε4 with the late-onset familial and sporadic Alzheimer disease. Neurology 43: 1467-1472

Schellenberg GD, Bird TD, Wijsman EM, Orr HT, Anderson L, Nemens E, White JA, Bonnycastle L (1992) Genetic linkage evidence for a familial Alzheimer's disease locus on chromosome 14. Science 258: 668-670

Scheuner D, Eckman L, Jensen M, Sung X, Citron M, Suzuki N, Bird T, Hardy J, Hutton M, Lannfelt L, Selkoe D, Younkin S (1996) Secreted amyloid-β protein similar to that in the senile plaques of Alzheimer disease is increased in vivo by presenilin 1 and 2 and APP mutations linked to FAD. Nature Med 2: 864-870

Sherrington R, Rogaev E, Liang Y, Rogaeva E, Levesque G, Ikeda M, Chi H, Lin C, Li G, Holman K, Tsuda T, Mar L, Foncin J-F, Bruni AC, Montesi MP, Sorbi S, Rainero I, Pinessi L, Nee L, Chumakov I, Pollen D, Brookes A, Sanseau P, Polinsky RJ, Wasco W, Da Silva MAR, Haines JL, Pericak-Vance MA, Tanzi RE, Roses AD, Fraser P, Rommens JM, St George-Hyslop P (1995) Cloning of a gene bearing missense mutations in early onset familial Alzheimer's disease. Nature 375: 754-760

Sherrington R, Froelich S, Sorbi S, Campion D, Chi H, Rogaeva EA, Levesque G, Rogaev EI, Lin C, Liang Y, Ikeda M, Mar L, Brice A, Agid Y, Percy ME, Clerget-Darpoux F, Karlinsky H, Piacentini S, Marcon G, Nacmias B, Amaducci L, Frebourg T, Lannfelt L, Rommens JM, St George-Hyslops PH (1996) Alzheimer's disease associated with mutations in presenilin 2 are rare and variably penetrant. Hum Molec Genet 5: 985-988

Sorbi S, Nacmias B, Forleo P, Amaducci L (1995) Epistatic effect of APP717 mutation and apoliproprotein E genotype in familial Alzheimer disease. Ann Neurol 38: 124-128

St George-Hyslop P, Haines J, Rogaev E, Mortilla M, Vaula G, Pericak-Vance M, Foncin J-F, Montesi M, Bruni A, Sorbi S, Rainero I, Pinessi I, Pollen D, Polinsky R, Nee L, Kennedy J, Macciardi F, Rogaeva E, Liang Y, Alexandrova N, Lukiw W, Schlumpf K, Tsuda T, Farrer L, Cantu J-M, Duara R, Amaducci L, Bergamini L, Gusella J, Roses A, Crapper MacLachlan D (1992) Genetic evidence for a novel familial Alzheimer disease gene on chromosome 14. Nature Genet 2: 330–334

St George-Hyslop PH, Tsuda T, Crapper McLachlan, Karlinsky H, Pollen D, Lippa C (1994) Alzheimer's disease and possible gene interaction. Science 263: 536–537

Tanzi RE, Vaula G, Romano D, Mortilla M, Huang T, Tupler R, Wasco W, St George-Hyslop P (1992) Assessment of amyloid β protein gene mutations in a large set of familial and sporadic Alzheimer disease cases. Am J Hum Genet 51: 273–282

The Alzheimer's disease Collaborative Group (1995) The structure of the presenilin 1 (S182) gene and the identification of six novel mutations in early onset AD pedigrees. Nature Genet 11: 219–222

Thinakaran G, Borchelt DR, Doan A, Slunt HH, Lee MK, Nordstedt C, Seeger M, Gandy SE, Hardy JA, Levey AI, Price DL, Sisodia SS (1996a) Endoproteolytic processing and protein topology of presenilin 1. Soc Neurosci 22: 728

Thinakaran G, Borchelt DR, Lee MK, Slunt HH, Spitzer L, Kim G, Ratovisky T, Davenport F, Nordstedt C, Seeger M, Levey AI, Gandy SE, Jenkins NA, Copeland N, Price DL, Sisodia SS (1996b) Endoproteolysis of presenilin 1 and accumulation of processed derivatives in vivo. Neuron 17: 181–190

Van Broeckhoven C, Backhovens H, Cruts M, De Winter G, Bruyland M, Cras P, Martin J-J (1992) Mapping of a gene predisposing to early-onset Alzheimer's disease to chromosome 14q24.3. Nature Genet 2: 335–339

Walter J, Capell A, Grunberg J, Pesold B, Schindzielorz A, Prior R, Podlisny MB, Fraser P, St George-Hyslop P, Selkoe D, Haass C (1996) The Alzheimer's disease associated presenilins are differentially phosphorylated proteins located predominantly within the endoplasmic reticulum. Mol Med 2: 673–691

Wong PC, Zheng H, Chen H, Trumbauer ME, Roskams AJ, Chen HY, Van der Ploeg LH, Price DL, Sisodia SS (1996) Functions of the presenilins: generation and characterization of presenilin 1 null mice. Soc Neurosc 22: 728

Alzheimer's Disease: A Matter of Dominance

*J. Hardy and M. Hutton**

The genetics of early onset autosomal dominant Alzheimer's disease has been very largely elucidated. A small proportion of disease is caused by mutations in the amyloid precursor protein gene (Goate et al. 1991), a moderate proportion by mutations in the presenilin 1 gene (Sherrington et al. 1995) and a small proportion by mutations in the presenilin 2 gene (Levy-Lahad et al. 1995). While there are many modest sized families with early onset disease in which no mutations have been found (Hutton et al. 1996), most of these are single sibships, which leaves open the possibility that, in these nuclear families, more complex genetic factors are responsible for disease. Our group had only two families that were larger than a single sibship in which we had not found a pathogenic mutation and, in at least one of these, we now believe that there is a structural PS1 mutation (see below). Thus it is possible that, in terms of simple pathogenic mutations → disease, we have discovered all the pathogenic loci. It is noteworthy that the nuclear families with early onset disease which remain in our collection have a high apolipoprotein E4 allele frequency, in contrast to the large kindreds in which we identified PS1 mutations, which independently suggests that more complex genetic factors are at work and that apolipoprotein E is one of these factors (authors' unpublished data).

Of particular and central importance is that mutations in all three pathogenic genes share the characteristic that, in a variety of tissues and preparations, they all lead to the overproduction of $A\beta 42(43)$ (Scheuner et al. 1996; Duff et al. 1996; Borchelt et al. 1996 see Duff, this volume, and Younkin, this volume) and this finding, together with the data implicating $A\beta 42(43)$ in amyloid fibrillogenesis (Jarrett et al. 1993) and in the pathology of the disease (Mann et al. 1996) has led to the "amyloid cascade hypothesis" becoming the dominant theory to explain the etiology and pathogenesis of the disease (for a recent review, see Hardy 1997).

The mechanisms of pathogenesis of mutations in the APP gene are well understood (except that the precise enzymes involved in APP processing are not known): they either potentiate β-secretase cleavage, inhibit α-secretase cleavage, or alter the position of γ-secretase cleavage (see Scheuner et al. 1996 and Hardy 1997 for references). By these three mechanisms, they lead to the production of

---

* Mayo Clinic Jacksonville, 4500 San Pablo Rd., Jacksonville, FL32224, USA

S. G. Younkin / R. E. Tanzi / Y. Christen (Eds.)
Presenilins and Alzheimer's Disease
© Springer-Verlag Berlin Heidelberg New York 1998

more Aβ42(43). However, the mechanisms that tie presenilin mutations to Aβ42(43) overproduction are not at all understood, and this problem is discussed here.

The putative structures of presenilin 1 and 2, together with the exon boundaries, the positions of mutations and the homologies between the proteins are shown in Figures 1 and 2. It is clear that the presenilin proteins are membrane proteins, and that in transfected cells they occur in Golgi and endoplasmic reticulum (Kovacs et al. 1996); however, they may also occur on other membranes. It is likely that both the N- and C-terminals of the proteins are cytoplasmic.

All the reported mutations, except one, are missense mutations (for review see Hardy 1997). This leads to the suggestion that the mutations are unlikely to be simple "loss of function" mutations since simple loss of function would most simply be caused by nonsense and frameshift mutations that have not yet been found. The single exception is the PS1 Δ9 mutation (Perez-Tur et al. 1996), in which exon 9 is deleted in frame through a splice site mutation. This mutation is of particular importance in considering the mechanisms of pathogenesis (see below).

The normal functions of the presenilins are not known; the clearest data comes from analysis of *sel12* and *spe4*, which are *C. elegans* homologues of the presenilins. It seems most likely that these are involved in Notch signalling and in protein trafficking in the Golgi, respectively. Both presenilin 1 and presenilin 2 can rescue the *sel12* phenotype in *C. elegans* (Levitan et al. 1996; Baumeister et al. 1997; see Haass et al., this volume). Interestingly, the many missense mutations in the presenilin 1 gene fail to rescue the phenotype effectively *(ibid)*. This suggests that there is a loss of function component to the mutations (see below). Surprisingly, however, the Δ9 mutation does rescue the *sel12* phenotype (Levitan et al. 1996; Baumeister et al. 1997; see Haass et al., this volume).

What is the Mode of Action of Autosomal Dominant Mutations Leading to AD?

With respect to the APP mutations, it is now clear that the mode of action of the mutations is a gain of a normal function of the APP molecule. Aβ42(43) appears to be a normal product of APP metabolism; however, the pathogenic mutations in APP appear to accentuate this metabolic route. Thus the mutations are „gain in function" mutations. To some extent, the presenilin mutations are similar in that they result in the same gain in function. However, from a molecular perspective, it is not immediately clear whether the presenilin mutations act as gain, loss or change of function mutations.

1) Loss of Function
It is unlikely that presenilin mutations are simple loss of function mutations because the simplest way to lose function would be through a nonsense or frame-

Fig. 1. Diagram showing the likely structure of presenilin 1, with exon boundaries and mutations illustrated (Clark et al. 1995; Hardy 1997)

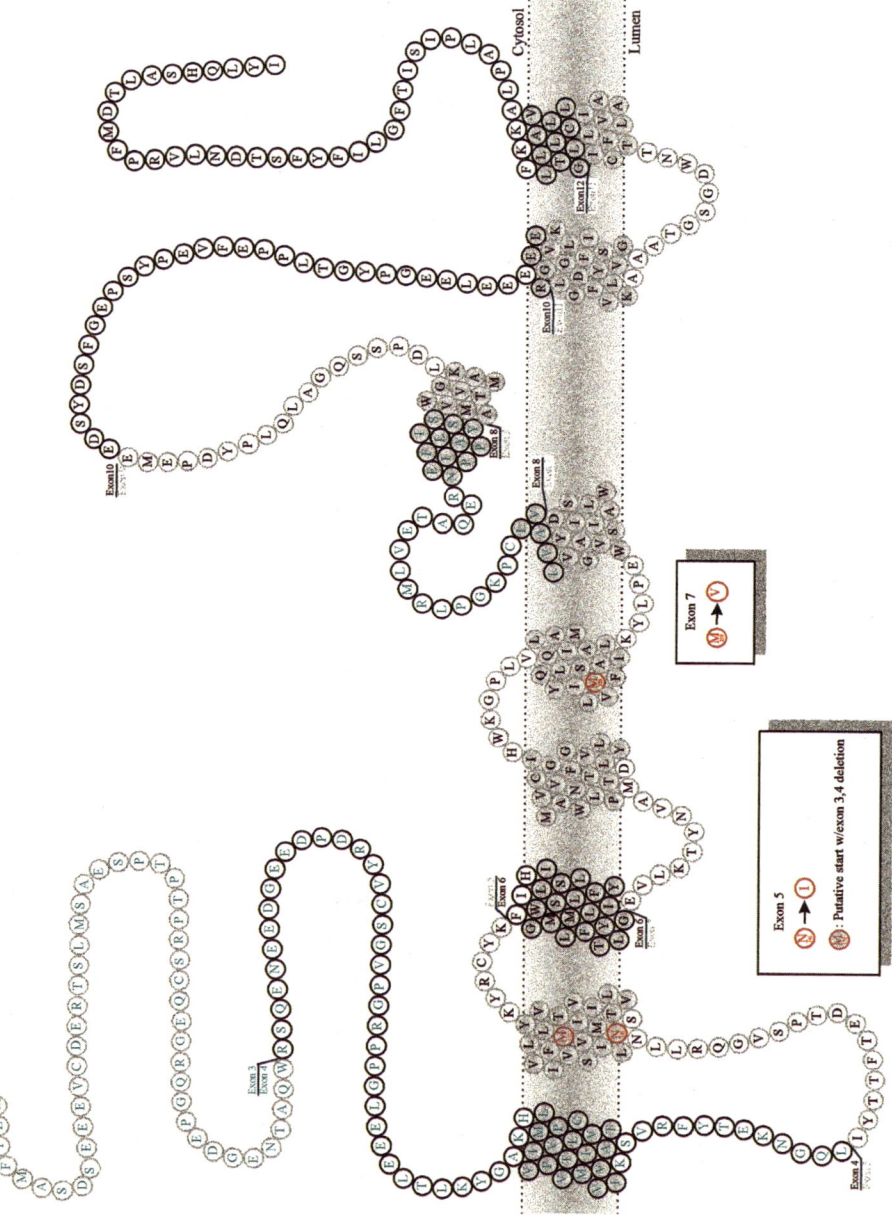

Fig. 2. Diagram showing the likely structure of presenilin 2, with exon boundaries and mutations, as well as the proposed start site of translation of those transcripts in which exons 3 and 4 are spliced out (Prihar et al. 1996; Haass and the authors' unpublished data)

shift change leading to the production of a non-functional protein. All the mutations are „careful" to maintain the integrity of the presenilin protein. On the other hand, the analogy with *sel12* suggests that there is likely to be a loss of function component to the mechanism of the mutation, although these data are derived from an apparently independent egg-laying phenotype. However, the fact that antisense-transfection protocols also lead to an Aβ42 cellular phenotype also suggest that a loss of function component may be important in the pathogenesis of Alzheimer's disease (Younkin, this volume).

2) Gain of Function
While the overall effect of presenilin mutations is a gain in function, unless the analogy with *sel12* is completely misleading, there must be a loss of function component to the presenilin mutations since the mutant proteins fail to compensate for the *sel12* knockout phenotype.

3) Dominant Negative
A dominant negative phenotype is one in which the mutant presenilin disrupts the function of the wild-type allele. This would imply a direct interaction between presenilin molecules. One could easily imagine interactions between different presenilin molecules in a hetero- or homo-multimeric structure. There is, however, no evidence for this idea.

4) Displacement
Displacement could be envisaged in a scheme in which there were only a limited number of sites in the cell at which presenilins could have their function, and the mutant presenilins could out-compete the wild-type presenilins at those sites (either for some structural reason or because they have a longer half life), but then be less effective at those sites. This hypothesis is difficult to distinguish from "dominant negative," but one might expect that cells or transgenic animals carrying mutant alleles would have altered amounts of wild-type protein.

The Δ9 Mutation

It is clear that most mutations in the presenilins are missense mutations. The sole exception, so far, has been the Δ9 mutation (Perez-Tur et al. 1996), which is a splice site mutation that cuts out exon 9 in frame from the protein. This mutation had been reported in two families so far (Perez-Tur et al. 1996; Kwok et al. 1997). In both of these families there are interesting, and slightly anomalous, data concerning the phenotype of the mutation. In the first family, while the clinical features are typical of Alzheimer's disease, the detailed pathological description reports rather anomalous Aβ deposition (Mann et al. 1996). In the second, the clinical phenotype begins with the very unusual presentation of paraplegia (Kwok et al. 1997; see also below).

A major processing pathway for presenilin 1 involves cleavage within this exon; the Δ9 mutation completely prevents this processing (Thinakaran et al. 1996). At face value, this observation is important for two reasons:
1) it suggests that full length protein, rather than the processed fragment is important in pathogenesis.
2) it suggests that inhibition of processing is one possible mechanism of pathogenesis. Interestingly, while the data are less strong, there is evidence that other presenilin mutations do alter presenilin processing (Mercken et al. 1996; Guo et al. 1996; Kim et al. 1997). Of course, inhibition of processing would fit well with the notion of displacement as a general mechanism of mutation pathogenesis (see above).

However, the Δ9 mutation may be anomalous in mechanism. There are three disparate but flawed pieces of data which point to this suggestion:
1) the Δ9 mutant rescues the *sel12* phenotype but other mutations do not.
2) the Δ9 mutant has the largest effect on Aβ42(43) production although the age of onset of disease in these families is not exceptional (author's unpublished data).
3) the Δ9 mutation appears, in some circumstances, to have unusual clinical presentations.

Other Structural Mutations in PS1?

Within our laboratory we have found ~ 15 families with PS1 missense mutations, a single family with the Δ9 PS1 mutation, five families with APP mutations and no families with PS2 mutations. We have three extended families (i.e., more than a single sibship with constant and early onset) in which we have not found APP or PS2 mutations (eliminated by linkage analysis as well as by limited sequencing) and in which we have not been able to find PS1 mutations. In two of these families we have enough samples for effective linkage analysis, and in both of these the marker D14S77 (the closest marker to PS1) shows evidence for segregation with disease. In one of these families, through analysis of cDNA from affected individuals, we have recently found evidence for a structural mutation deleting part of the gene. Thus it remains possible that there are other structural mutations, besides the Δ9 exon splice mutation, within PS1 which can lead to disease (our basic mutation searching strategy will pick up all missense mutations and mutations that occur within ~ 20 bp of an exon; Hutton et al. 1996). Thus, there are no families that consist of more than a single sibship within our dataset which are unequivocally not caused by APP, PS1 or PS2 mutations. In this regard, we have previously described our ascertainment procedure which selected for families with early and constant onset age (Hutton et al. 1996).

In addition, we have many families within our dataset which consist of single sibships with early onset in which we have not found mutations, but these families have a high apolipoprotein E4 allele frequency, leaving open the possibility that this form of the disease is polygenic (explaining why they are single sibships) with apolipoprotein E as one of the genes.

References

Baumeister R, Leimer U, Zweckbronner I, Jakubek C, Grunberg J, Haass C (1997) Proteolytic cleavage of the Alzheimer's disease associated presenilin 1 is not required for its function in C. elegans. Genes Function 1997, in press

Borchelt DR, Thinakaran G, Eckman CB, Lee MK, Davenport F, Ratovisky T, Prada CM, Kim G, Seekins S, Yager D, Slunt HH, Wang R, Seeger M, Levey M, Levey AI, Gandy SE, Copeland NG, Jenkins NA, Price DL, Younkin SG, Sisodia SS (1996) Familial Alzheimer's disease linked presenilin 1 variants elevate $A\beta 1$-42/1-40 ratio *in vitro* and *in vivo*. Neuron 17: 1005–1013

Clark RF, Hutton M, Fuldner RA, Froelich S, Karran E, Talbot C, Crook R, Lendon C, Prihar G, He C, Korenblat K, Martinez A, Wragg M, Busfield F, Behrens MI, Myers A, Norton J, Morris J, Mehta N, Pearson C, Lincoln S, Baker M, Duff K, Zehr C, Perez-Tur J, Houlden H, Ruiz A, Ossa J, Lopera F, Arcos M, Madrigal L, Collinge J, Humphreys C, Ashworth A, Sarner S, Fox N, Harvey R, Kennedy A, Roques P, Cline RT, Philips CA, Venter JC, Forsell L, Axelman K, Lilius L, Johnston J, Cowburn R, Viitanen M, Winblad B, Kosik K, Haltia M, Poyhonen M, Dickson D, Mann D, Neary D, Snowden J, Lantos P, Lannfelt L, Rossor M, Roberts GW, Adams MD, Hardy J, Goate A (1995) The structure of the presenilin 1 (S182) gene and identification of six novel mutations in early onset AD families. Nature Genet 11: 219–222

Duff K, Eckman C, Zehr C, Yu X, Prada CM, Perez-Tur J, Hutton M, Buee L, Harigaya Y, Yager D, Morgan D, Gordon MN, Holcomb L, Refolo L, Zenk B, Hardy J, Younkin S (1996) Increased amyloid-$\beta 42(43)$ in brains of mice expressing mutant presenilin 1. Nature 383: 710–713

Goate AM, Chartier-Harlin MC, Mullan MC, Brown J, Crawford F, Fidani L, Giuffra L, Haynes A, Irving N, James L, Mant R, Newton P, Rooke K, Roques P, Talbot C, Pericak-Vance M, Roses A, Williamson R, Rossor MN, Owen M, Hardy J (1991) Segregation of a missense mutation in the amyloid precursor protein gene with familial Alzheimer's disease. Nature 349: 704–706

Guo Q, Sopher BL, Pham DG, Robinson N, Martin GM, Mattson MP (1996) Alzheimer's PS1 mutation perturbs calcium homeostasis and sensitizes PC12 cells to death induced by amyloid β-peptide NeuroReport 8: 379–383

Hardy J (1997) Amyloid, the presenilins and Alzheimer's disease. Trends Neurosci 20: 154–159

Hutton M, Busfield F, Wragg M, Crook R, PerezTur J, Clark RF, Prihar G, Talbot C, Phillips H, Wright K, Baker M, Lendon C, Duff K, Martinez A, Houlden H, Nichols A, Karran E, Roberts G, Roques P, Rossor M, Venter JC, Adams MD, Cline RT, Phillips CA, Fuldner RA, Hardy J, Goate A (1996) Complete analysis of the presenilin 1 gene in families with early onset Alzheimer's disease. NeuroReport 7: 801–805

Jarrett JT, Berger EP, Lansbury PT (1993) The carboxy terminus of β-amyloid protein is critical for the seeding of amyloid formation: implications for the pathogenesis of Alzheimer's disease. Biochemistry 32: 4693–4697

Kim T-W, Hallmark OG, Pettingell WH, Wasco W, Tanzi RE (1997) Proteolytic processing and ubiquitin-proteasomal, degradation of wild type and mutant forms of presenilin 2. In: Iqbal K, Winblad B, Nishimura T, Takeda M, Wisniewski H (eds) Alzheimer's disease: biology, diagnosis and therapeutics. John Wiley, Chichester, pp 575–582

Kovacs DM, Fausett HF, Page KJ, Kim TW, Moir RD, Merriam DE, Hollister RD, Hallmark OG, Mancini R, Felsenstein KM, Hyman B, Tanzi RE, Wasco W (1996) Alzheimer associated presenilin 1 and 2: neuronal expression in brain and localization to intracellular membranes in mammalian cells. Nature Med 2: 224–229

Kwok JBJ, Taddel K, Hallupp M, Fisher C, Brook WS, Broe G, Hardy J, Fulham MJ, Nicholson GA, Stell R, St George Hyslop PH, Fraser PE, Kakulas B, Clarnette R, Relkin N, Gandy SE, Schofield PR, Martins RN (1997) Two novel (M233T and R278T) presenilin-1 mutations in early onset Alzheimer's disease and preliminary evidence for association of presenilin-1 mutations with a novel phenotype. Neuro-Report, in press

Levitan DH, Doyle TG, Brousseau D, Lee MK, Thinakaran G, Shunt HH, Sisodia SS, Greenwald I (1996) Assessment of normal and mutant human presenilin function in Caenorhabditis elegans. Proc Natl Acad Sci USA 93: 14940–14944

Levey-Lahad E, Wasco W, Poorkaj P, Romano DM, Oshima J, Pettingell WH, Yu CE, Jondo PD, Schmidt SD, Wang K, Crowley AC, Fu YH, Guenette SY, Galas D, Nemens E, Wijsman EM, Bird TD, Schellenberg GD, Tanzi RE (1995) Candidate gene for the chromosome 1 familial Alzheimer's disease locus. Science 269: 973–977

Mann DMA, Iwatsubo T, Cairns NJ, Lantos PL, Nochlin D, Sumi SM, Bird TD, Poorkaj P, Hardy J, Hutton M, Prihar G, Crook R, Rossor MN, Haltia M (1996) Amyloid β protein (Aβ) deposition in chromosome 14-linked Alzheimer's disease: predominance of Aβ42(43). Ann Neurol 40: 149–156

Mercken M, Takahashi H, Honda T, Sato K, Murayama M, Nakazato Y, Noguchi K, Imahori K, Takashima A (1996) Characterization of human presenilin 1 using N-terminal specific monoclonal antibodies: evidence that Alzheimer mutations affect proteolytic processing. FEBS Lett 389: 297–303

Perez-Tur J, Froelich S, Prihar G, Crook R, Baker M, Duff K, Wragg M, Busfield F, Lendon C, Clark RF, Roques P, Fuldner RA, Johnston J, Cowburn R, Forsell C, Axelman K, Lilius L, Houlden H, Karran E, Goate A, Lannfelt L, Hutton M (1995) A mutation in Alzheimer's disease destroying a splice acceptor site in the presenilin 1 gene. NeuroReport 7: 297–301

Prihar G, Fuldner R, Perez-Tur J, Lincoln S, Duff K, Crook R, Hardy J, Philips CA, Venter C, Talbot C, Clark RF, Goate A, Li J, Potter H, Karran E, Roberts GW, Hutton M, Adams MD (1996) The structure and alternate splicing of the presenilin 2 gene. NeuroReport 7: 1680–1684

Scheuner D, Eckman C, Jensen M, Song X, Citron M, Suzuki N, Bird TD, Hardy J, Hutton M, Kukull W, Larson E, Levey-Lahad E, Viitanen M, Peskind E, Poorkaj P, Schellenberg G, Tanzi R, Wasco W, Lannfelt L, Selkoe D, Younkin S (1996) Secreted amyloid β-protein similar to that in the senile plaques of Alzheimer's disease is increased in vivo by the presenilin 1 and 2 and APP mutations linked to familial Alzheimer's disease. Nature Med 2: 864–870

Sherrington R, Rogaev EI, Liang Y, Rogaeva EA, Levesque G, Ikeda M, Chi H, Lin C, Li G, Homan K, Tsuda T, Mar L, Foncin JF, Bruni AC, Montesi MP, Sorbi S, Rainero I, Pinessi L, Nee L, Chumakov I, Pollen D, Brookes A, Sanseau P, Polinsky RJ, Wasco W, Da Silva HAR, Haines JL, Pericak-Vance MA, Tanzi RE, Roses AD, Fraser PE, Rommens JM, St George Hyslop PH (1995) Cloning of a gene bearing missense mutations in early onset familial Alzheimer's disease. Nature 375: 754–760

Thinakaran G, Borchelt DR, Lee MK, Slunt HH, Spitzer L, Kim G, Ratovitsky T, Davenport F, Nordstedt C, Seeger M, Hardy J, Levey AI, Gandy SE, Jenkins NA, Copeland NG, Price DL, Sisodia S (1996) Endoproteolysis of presenilin 1 and accumulation of processed derivatives in vivo. Neuron 17: 181–190

Alternative Endoproteolysis of the Presenilins and Familial Alzheimer's Disease

R. E. Tanzi*[*], T.-W. Kim, D. M. Kovacs, W. Wasco, R. D. Moir, W. Pettingell, J. Henderson and R. Mancini

Introduction

Familial Alzheimer's disease (FAD) is a genetically heterogeneous neurodegenerative disorder which can be caused by defects in at least three early-onset (< 60 yr.) genes located on chromosomes 1 (presenilin 2, PS2; Levy-Lahad et al. 1995), 14 (Presenilin 1, PS1; Sherrington et al. 1995) and 21 (amyloid β protein precursor, APP; Tanzi et al. 1987; Goate et al. 1991). The enhanced deposition of Aβ and formation of β-amyloid appear to be a common pathogenic phenotype associated with all of the known early-onset FAD genes (Citron et al. 1992; Suzuki et al. 1994). Increased production of β-amyloid has also been observed in middle-aged patients with Down Syndrome (DS, trisomy 21), most likely due to the presence of three copies of the gene encoding the amyloid β-protein precursor on chromosome 21. Mutations within or immediately proximal to the Aβ portion of the APP gene are responsible for only a small proportion (2–3%) of reported cases of FAD (Tanzi et al. 1992), whereas up to half of all cases of early onset FAD are caused by mutations in the PS1 and PS2 genes (reviewed in Tanzi et al. 1996a,b).

Plasma and fibroblasts from patients and at-risk carriers with the presenilin mutations have been shown to contain increased amounts of a longer, more amyloidogenic form of the Aβ peptide, Aβx-42 (Scheuner et al. 1996). Similar increases have been observed in transfected cell lines and transgenic animals expressing mutant forms of PS1 and PS2 (Borchelt et al. 1996; Duff et al. 1996; Citron et al. 1997; Tomita et al. 1997). Increased deposition of Aβ42 (Lemere et al. 1996; Gomez-Isla et al. 1997) and Aβ40 (Gomez-Isla et al. 1997) has been observed in the brains of patients with PS1 FAD mutations. The critical role for Aβ deposition in AD is further strengthened by the fact that the apolipoprotein E4 (APOE4) isoform of APOE (chromosome 19) confers increased risk for late onset FAD, especially in cases with onset between 60 and 70 years (Saunders et al. 1993; Blacker et al. 1997), and is also associated with significantly increased amyloid burden in AD and DS patients who are APOE4-positive relative to those patients who do not carry an APOE4 allele (Hyman et al. 1995). Collectively, the

---

[*] Genetics and Aging Unit and the Department of Neurology, Massachusetts General Hospital, Harvard Medical School, 149, 13th St. Charlestown, MA 012129

genetic and corresponding phenotypic data support the notion that mutations in APP and the presenilins influence the processing of APP (particularly at the gamma-secretase site allowing for increased generation of Aβ42, the longer more amyloidogenic form of Aβ), whereas APOE most likely influences the rate of Aβ aggregation and subsequent β-amyloid deposition (Tanzi et al. 1996a,b).

The Presenilins

The presenilin proteins exhibit a "serpentine" topology with six to nine predicted transmembrane (TM) domains (Sherrington et al. 1995; Levy-Lahad et al. 1995; seven TM; Slunt et al. 1995; nine TM; Doan et al. 1996; Li and Greenwald 1996; eight TM; Lehmann et al. 1997; six TM). Although PS1 and PS2 share 67% amino acid identity, two particularly non-homologous regions are found in the N-terminus and the large hydrophilic loop lying between predicted TM6 and TM7. PS1 and PS2 both appear to be ubiquitously expressed throughout the body (Sherrington et al. 1995; Rogaev et al. 1995; Levy-Lahad et al. 1995) and *in situ* hybridization analysis for rat brain has shown that PS1 is expressed predominantly in neurons (Kovacs et al. 1996; Page et al. 1996). In the brain, the highest levels of PS1 are found in the hippocampus, cerebellum, and choroid plexus followed by cortex, striatum, and midbrain. Expression of PS1 in glia in the white matter is relatively low.

Both PS1 and PS2 are primarily localized to intracellular membranes in the endoplasmic reticulum (ER) and, to a lesser extent, the Golgi (Kovacs et al. 1996). Meanwhile, no staining for PS1 or PS2 was observed on the plasma membrane. Cook et al. (1996) have recently shown PS1 to be localized to the rough ER as well as in dendrites of NT2N human neuronal cells infected with human PS1. The biological role of the presenilins is not known. However, clues can be derived from the fact that both PS1 and PS2 share homology with two *C. elegans* proteins, *sel-12* (Levitan and Greenwald 1995) and *spe-4* (L'Hernault and Arduengo 1992). SEL-12 may function as a co-receptor for LIN-12, facilitate downstream in the LIN-12 Notch receptor signaling pathway, or play a role in the trafficking or recycling of LIN-12. Mutations in *spe-4* have been shown to disrupt the fibrous body membrane organelle (FB-MO) which, during spermatogenesis in the nematode, transports proteins designated for degradation in the residual body. Together, these homologies, along with the subcellular localization of the presenilins to the ER and Golgi, suggest that the presenilins may play roles in intracellular trafficking and transport. Since FAD mutations in both PS1 and PS2 lead to relative increases in the ratio of Aβ42:Aβ40, it is reasonable to assume that the normal intracellular trafficking and/or processing of APP is disrupted either directly or indirectly by presenilin FAD mutations.

Presenilin Processing and Metabolism

To begin to understand the biology of the presenilins, we have recently analyzed and compared the processing and degradation of PS1 and PS2 in inducible cell lines expressing these proteins. Full-length PS2 was observed in the transfected cells as a 53–54 kDa band (Kim et al. 1997a,b). In H4 cells stably transfected with PS1, the full-length species was observed as a 48 kDa band, in general agreement with published reports (Thinakaran et al. 1996; Duff et al. 1996; Citron et al. 1997; Mercken et al. 1996), and for both PS1 and PS2, the full-length endogenous species were largely undetectable in brain and native H4 human neuroglioma cells. However, some antisera are apparently able to recognize full-length endogenous forms of the presenilins (S. Younkin, personal communication). Both of presenilins undergo regulated proteolytic processing to yield two stable fragments. In native H4 cells, we have observed the endogenous endoproteolytic fragments of PS1 to be 27 kDa (N-terminal) and 21 kDa (C-terminal); these fragments have previously been shown to be both saturable and regulated (Thinakaran et al. 1996). Along similar lines, we have found that PS2 undergoes regulated endoproteolytic cleavage in brain and native H4 cells to generate a 30 kDa N-terminal fragment and a 25 kDa C-terminal fragment (unpublished observations).

Recently, we have found that when PS2 is overexpressed in stably transfected, inducible H4 cells, an alternative, non-regulated endoproteolytic clip occurs distal of the normal cleavage site to yield a 34 kDa N-terminal and 20 kDa C-terminal fragment (Kim et al. 1997a,b; Tanzi et al. 1996a,b). The 20 kDa alternative PS2-CTF is enriched in the detergent (1% triton)-resistant cellular fraction and exhibits a very slow rate of turnover (Kim et al. 1997a,b; Tanzi et al. 1996a,b). Meanwhile, the normal regulated 25 kDa C-terminal fragment is found only in the detergent-soluble fraction. The localization of the 20 kDa alternative PS2 C-terminal fragment to the detergent-resistant fraction suggests that it may be associated with cytoskeletal elements. Alternatively, this fragment may be able to self-aggregate into a triton-resistant protein complex/aggregate.

We have previously shown that PS2 is poly-ubiquitinated when overexpressed in the transfected H4 cell lines and is degraded by the ubiquitin-proteasome pathway (Kim et al. 1997a,b; Tanzi et al. 1996a,b). It is therefore conceivable that proteasomal degradation of full-length PS2 serves as a means for regulating the endoproteolysis of the full-length protein. Interestingly, when the proteasome was blocked with either ALLN or lactacystin, an elevation in ubiquitinated, high-molecular weight PS2 material and the alternative C-terminal fragment occurred. Overexpression of PS2 and inhibition of proteasomal degradation lead to not only an increase in the alternatively cleaved PS2 fragments, but also in the amount of full-length PS2. Thus, we postulate that excess levels of full-length PS2 may override the regulated endoproteolytic cleavage pathway and divert PS2 into the alternative endoproteolytic pathway which yields the detergent-resistant C-terminal fragment. The alternative C-terminal fragment has also been shown to be localized exclusively in the ER, whereas the normal detergent-soluble C-terminal fragment is found in both the ER and Golgi (W. Wasco, personal communication).

We have observed a processing/degradation pathway for PS1 that is similar to that for PS2. When PS1 was highly overexpressed in stably-transfected H4 cells, the normal regulated 27 kDa N-terminal and 21 kDa C-terminal fragments along with two alternative cleavage fragments were generated. The alternative fragments were 34 kDa (N-terminal) and 14 kDa (C-terminal). As was observed for PS2, the alternative 14 kDa C-terminal fragment, but not the regulated 21 kDa C-terminal fragment, was enriched in the detergent-resistant fraction. Moreover, inhibition of PS1 degradation by the proteasome using lactacystin or ALLN led to an increase in ubiquitinated high molecular weight forms of PS1 and elevated amounts of the alternative fragments.

Collectively, our data on the processing and degradation of the presenilins indicate that increases in the level of full-length PS1 or PS2 divert the presenilins into alternative cleavage pathways and lead to the generation of highly stable, detergent-resistant C-terminal fragments. Increased levels of full-length PS1 or PS2 can be attained by either overexpression or inhibition of proteasome degradation of endogenous or transgene-derived PS1/PS2 (Fig. 1). But the most

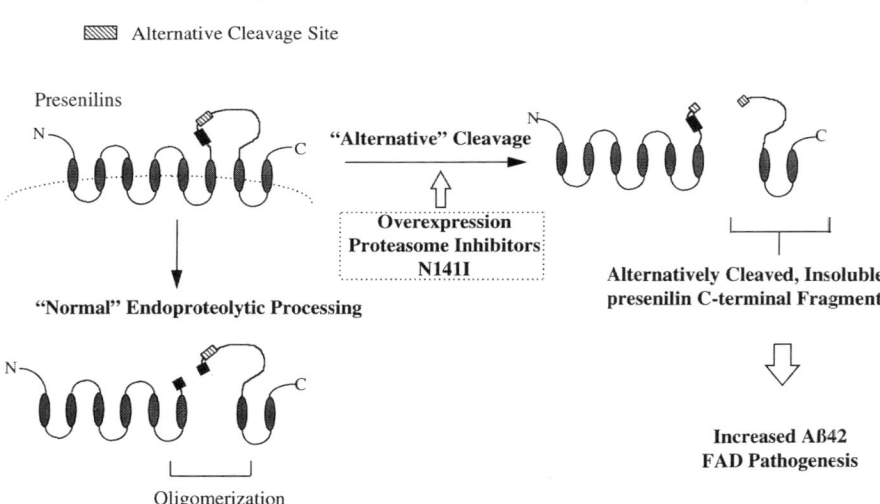

Fig. 1. The normal and alternative cleavage pathways of the presenilins. The "Normal" endoproteolytic pathway is regulated leading to a saturable amount of N-terminal and C-terminal fragments. The normal clip site is encoded in exon nine of PS1 and PS2. Following their generation, the two detergent-soluble cleavage products can undergo oligomerization. When PS1 or PS2 are overexpressed or their degradation is blocked with proteasomal inhibitors, the full-length presenilins can be diverted into an alternative pathway in which cleavage takes place distal of the normal endoproteolytic clip site and gives rise to a larger N-terminal fragment and a smaller, detergent-resistant (insoluble) C-terminal fragment. Since the Volga German FAD mutation PS2-N141I leads to increased ratios of Aβ42:Aβ40, it is postulated that the generation of the alternative presenilin cleavage fragments may contribute to abnormal processing of APP and FAD pathogenesis

interesting finding is that the Volga German N141I FAD mutation can directly lead to excess generation of the alternative cleavage fragments.

Conclusions

These data beg the question of whether the generation of the alternative presenilin endoproteolytic fragments enhances, compromises, or has no effect on cell viability. For PS2, we have observed that PS2 containing the Volga German FAD mutation N141I leads to a 3.5-fold increase in the ratio of alternative (20 kDa): normal (25 kDa) C-terminal fragments in a comparison of stably transfected inducible cells lines expressing either wild-type PS2 or PS2-N141I (five clonal cell lines each). These observations imply that the generation of the alternative PS2 endoproteolytic fragments may be directly involved in the neuropathogenic mechanism of the PS2-N141I mutation. It is interesting to note that the splice acceptor FAD mutation that leads to the deletion of exon nine in PS1 (Perez-Tur et al. 1996) would also remove the normal cleavage site while preserving the more distal alternative clip site. In preliminary studies, we have also observed that the rate of production of the alternative detergent-resistant PS2 C-terminal fragment correlates with the degree of apoptotic cell death that is observed over time in the H4 cells induced to express PS2-N141I (unpublished observations). Moreover, other preliminary studies reveal that prevention of the alternative cleavage of the FAD mutant PS2-N141I appears to reduce the increase in the Aβ42:Aβ40 ratio associated with this mutation (unpublished observations). It is thus important in future studies to test whether therapeutic strategies aimed at blocking the alternative cleavage of the presenilins can effectively and consistently prevent apoptotic cell death or increases in the Aβ42:Aβ40 ratio that are observed with FAD mutant forms of the presenilins.

Acknowledgments. This work was supported by grants from the NIA, NINDS, and The Metropolitan Life Foundation. D. Kovacs is a French Foundation Fellow.

References

Blacker D, Haines JL, Rodes L, Terwedow H, Go R, Harrell L, Perry R, Bassett SS, Chase G, Meyers D, Albert M, Tanzi RE (1997) APOE-4 and age of onset of Alzheimer's disease: The NIMH Genetics Initiative. Neurology 48: 139–147

Borchelt DR, Thinakaran G, Eckman CB, Lee MK, Davenport F, Ratovitsky T, Prada C-M, Kim G, Seekins S, Yager D, Slunt HH, Wang R, Seeger M, Levey AI, Gandy SE, Copeland NG, Jenkins NA, Price DL, Younkin SG, Sisodia SS (1996) Familial Alzheimer's disease-linked presenilin 1 variants elevate Aβ-42/1-40 ratio in vitro and in vivo. Neuron 17: 1005–1013

Citron M, Oltersdorf T, Haass C, McConlogue L, Hung AY, Seubert P, Vigo-Pelfrey C, Lieberburg I, Selkoe DJ (1992) Mutation of the β-amyloid precursor protein in familial Alzheimer's disease increases β-protein production. Nature 360: 672–674

Citron M, Westaway D, Xia W, Carlson G, Diehl T, Levesque G, Johnson-Wood K, Lee M, Seubert P, Davis A, Kholodenko D, Motter R, Sherrington R, Perry B, Yao H, Strome R, Lieberburg I, Rommens J, Kim S, Schenk D, Fraser P, St George-Hyslop PS, Selkoe DJ (1997) Mutant presenilins of Alzheimer's disease increase production of 42-residue amyloid beta-protein in both transfected cells and transgenic mice. Nat Med 3: 67–72

Cook DG, Sung JC, Golde TE, Felsenstein KM, Wojczk BS, Tanzi RE, Trojanowski JQ, V M-Y Lee, Doms RW (1996) Expression and analysis of presenilin 1 in a human neuronal system: localization in cell bodies and dendrites. Proc Natl Acad Sci USA 93: 9223–9228

Doan A, Thinakaran G, Borchelt DR, Slunt HH, Ratovitsky T, Podlisny M, Selkoe DJ, Seeger M, Gandy SE, Price DL, Sisodia SS (1996) Protein topology of presenilin 1. Neuron 17: 1023–1030

Duff K, Eckman C, Zehr C, Yu X, Prada C-M, Perez-Tur J, Hutton M, Buee L, Harigaya Y, Yager D, Morgan D, Gordon MN, Holcomb L, Refolo L, Zenk B, Hardy J, Younkin S (1996) Increased amyloid-$\beta 42(43)$ in brains of mice expressing mutant presenilin 1. Nature 383: 710–713

Goate A, Chartier-Harlin M, Mullan M, Brown J, Crawford F, Fidani L, Giuffra L, Haynes A, Irving N, James L, Mant R, Newton P, Rooke K, Roques P, Talbot C, Pericak-Vance M, Roses A, Williamson R, Rossor M, Owen M, Hardy J (1991) Segregation of a missense mutation in the amyloid precursor protein gene with familial Alzheimer's disease. Nature 349: 704–706

Gomez-Isla T, Wasco W, Pettingell WP, Garubhagavatula S, Schmidt DD, Jondro PD, McNamara M, Rodes LA, DiBlasi T, Growdon WB, Seubert P, Schenk D, Growdon JH, Hyman B, Tanzi RE (1997) Novel Presenilin 1 gene mutation: Increased β-amyloid and neurofibrillary changes. Ann Neurol 41: 809–813

Hyman BT, West HL, Rebeck GW, Buldyrev SV, Mantegna RN, Ukleja M, Havlin S, Stanley HE (1995) Quantitative analysis of senile plaques in Alzheimer's disease: Observation of log-normal size distribution and molecular epidemiology of differences associated with ApoE genotype and trisomy 21 (Down syndrome). Proc Natl Acad Sci USA 92: 3586–3590

Kim T-W, Hallmark OG, Pettingell W, Wasco W, Tanzi RE (1997a) Proteolytic processing and Ubiquitin-proteasomal degradation of wild-type and mutant forms of presenilin 2. In: Iqbal K, Winblad B, Nishimura T, Takeda M, Wisniewski (eds) Alzheimer's disease: biology, diagnosis and therapeutics. John Wiley, Chichester, pp 575–582

Kim T-W, Hallmark OG, Pettingell WH, Wasco W, Tanzi RE (1997b) Endoproteolytic cleavage and proteasomal degradation of presenilin 2 in transfected cells. J Biol Chem 272: 1106–1110

Kovacs DM, Fausett HJ, Page KJ, Kim T-W, Mori RD, Merriam DE, Hoillister RD, Hallmark OG, Mancini R, Felsenstein KM, Hyman BT, Tanzi RE, Wasco W (1996) Alzheimer associated presenilins 1 and 2: neuronal expression in brain and localization to intracellular membranes in mammalian cells. Nat Med 2: 224–229

Lehmann S, Chiesa R, Harris DA (1997) Evidence for a six-transmembrane domain structure of presenilin 1. J Biol Chem 272: 12047–12051

Lemere CA, Lopera F, Kosik KS, Lendon CL, Ossa J, Saido TC, Yamaguchi H, Ruiz A, Martinez A, Madrigal L, Hincabie L, Arango JC, Anthony DC, Koo EH, Goate AM, Selkoe DJ, Arango JCV (1996) The E280A presenilin 1 Alzheimer mutation produces increased A$\beta 42$ deposition and severe cerebellar pathology. Nat Med 2: 1146–1150

Levitan D, Greenwald I (1995) Facilitation of *lin-12*-mediated signalling by *sel12*, a *Caenorhabditis elegans* S182 Alzheimer's disease gene. Nature 377: 351–354

Levy-Lahad E*, Wasco W*, Poorkaj P, Romano DM, Oshima JM, Pettingell WH, Yu C, Jondro PD, Schmidt SD, Wang K, Crowley AC, Fu Y-H, Guenette SY, Galas D, Nemens E, Wijsman EM, Bird TD, Schellenberg GD, Tanzi RE (1995) Candidate gene for the chromosome 1 familial Alzheimer's disease locus. Science 269: 973–977. (*shared first author)

L'Hernault SW, Arduengo PM (1992) Mutation of a putative sperm membrane protein in *Caenorhabditis elegans* prevents sperm differentation but not its associated meiotic divisions. J Cell Biol 119: 55–68

Li X, Greenwald I (1996) Membrane topology of the *C. elegans* SEL-12 presenilin. Neuron 17: 1015–1021

Mercken M, Takahashi H, Honda T, Sato K, Murayama M, Nakazato Y, Noguchi K, Imahori K, Takashima A (1996) Characterization of human presenilin 1 using N-terminal specific monoclonal antibodies: evidence that Alzheimer mutations affect proteolytic processing. FEBS Lett 389: 297–303

Page K, Hollister R, Tanzi RE, Hyman BT (1996) In Situ hybridization of presenilin 1 mRNA in Alzheimer's disease and in lesioned rat brain. Proc Natl Acad Sci USA 93: 14020–14024

Perez-Tur J, Froelich S, Prihar G, Crook R, Baker M, Duff K, Wragg M, Busfield F, Lendon C, Clark RF, Roques P, Fuldner RA, Johnston J, Cowburn R, Forsell C, Axelman K, Lilius L, Houlden H, Karran E, Roberts GW, Rossor M, Adams MD, Hardy J, Goate A, Lannfelt L, Hutton M (1996) A mutation in Alzheimer's disease destroying a splice acceptor site in the presenilin-1 gene. Neuroreport 7: 297–301

Rogaev EI, Sherrington R, Rogaeva EA, Levesque G, Ikeda M, Liang Y, Chi H, Lin C, Holman K, Tsuda T, Mar L, Sorbi S, Nacmias B, Piacentini S, Amaducci L, Chumakov I, Cohen D, Lannfelt L, Fraser PE, Rommens JM, St George-Hyslop P (1995) Familial Alzheimer's disease in kindreds with missense mutation in a gene on chromosome 1 related to the Alzheimer's disease type 3 gene. Nature 376: 775–778

Saunders AM, Strittmatter WJ, Schmechel D, St George-Hyslop PH, Perikak-Vance MA, Joo SH, Rosi B, Gusella JF, Crapper-MacLachlan DR, Alberts MJ, Hulette C, Crain B, Goldgaber D, Roses AD (1993) Association of apolipoprotein E allele E4 with late-onset familial and sporadic Alzheimer's disease. Neurology 43: 1467–1472

Scheuner D, Eckman C, Jensen M, Song X, Citron M, Suzuki N, Bird TD, Hardy J, Hutton M, Kukull W, Larson E, Levy-Lahad E, Viitanen M, Peskind E, Poorkaj P, Schellenberg G, Tanzi RE, Wasco W, Lannfelt L, Selkoe D, Younkin S (1996) Aβ42(43) is increased in vivo by the PS1/2 and APP mutations linked to familial Alzheimer's disease. Nat Med 2: 864–870

Seeger M, Norstedt C, Petanceska S, Kovacs DM, Gouras GK, Hahne S, Fraser P, Levesque L, Czernik AJ, St. George-Hyslop P, Sisodia SS, Thinakaran G, Tanzi RE, Greengard P, Gandy S (1997) Evidence for phosphorylation and oligomeric assembly of presenilin 1. Proc Natl Acad Sci USA 94: 5090–5094

Sherrington R, Rogaev EI, Liang Y, Rogaeva EA, Levesque G, Ikeda M, Chi H, Lin C, Li G, Holman K, Tsuda T, Mar L, Foncin J-F, Bruni AC, Montesi MP, Sorbi S, Rainero I, Pinessi L, Nee L, Chumakov Y, Pollen D, Wasco W, Haines JL, Da Silva R, Pericak-Vance M, Tanzi RE, Roses AD, Fraser PE, Rommens JM, St George-Hyslop PH (1995) Cloning of a novel gene bearing missense mutations in early onset familial Alzheimer disease. Nature 375: 754–760

Slunt HH, Thinakaran G, Lee MK, Sisodia SS (1995) Nucleotide sequence of the chromosome 14-encoded S182 cDNA and revised secondary structure prediction. Amyloid Int J Exp Clin Invest 2: 188–190

Suzuki N, Cheung TT, Cai XD, Odaka A, Otvos L Jr, Eckman C, Golde TE, Younkin SG (1994) An increased percentage of long amyloid beta protein secreted by familial amyloid beta protein precursor (beta APP717) mutants. Science 264: 1336–1340

Tanzi RE, Gusella JF, Watkins PC, Bruns GAP, St George-Hyslop P, van Keuren ML, Patterson D, Pagan S, Kurnit DM, Neve RL (1987) Amyloid β protein gene: cDNA, mRNA distribution and genetic linkage near the Alzheimer locus. Science 235: 880–884

Tanzi RE, Vaula G, Romano DM, Mortilla M, Huang TL, Tupler RG, Wasco W, Hyman BT, Haines JL, Jenkins BJ, Kalaitsidaki M, Warren AC, McInnis MG, Antonarakis SE, Karlinsky H, Percy ME, Connor L, Growdon J, Crapper-McLachlan DR, Gusella JF, George-Hyslop PH (1992) Assessment of amyloid β protein precursor gene mutations in a large set of familial and sporadic Alzheimer disease cases. Am J Human Genet 51: 273–282

Tanzi RE, Kovacs DM, Kim T-W, Moir RD, Guenette SY, Wasco W (1996a) The gene defects responsible for familial Alzheimer's disease. Neurobiol Disease 3: 159–168

Tanzi RE, Kovacs DM, Kim T-W, Moir RD, Guenette SY, Wasco W (1996b) The presenilin Genes and their role in early-onset familial Alzheimer's disease. Alzheimer's Disease Rev 1: 91–98

Thinakaran G, Borchelt D, Lee M, Slunt H, Spitzer L, Kim G, Ratovitsky T, Davenport F, Nordstedt C, Seeger M, Hardy J, Levey AI, Gandy SE, Jenkins NA, Copeland NG, Price DL, Sisodia SS (1996) Endoproteolysis of presenilin 1 and accumulation of processed derivatives in vivo. Neuron 17: 181–190

Tomita T, Maruyama K, Saido TC, Kume H, Shinozaki K, Tokuhiro S, Capell A, Walter J, Gruenberg J, Haass C, Iwatsubo T, Obata K (1997) The presenilin 2 mutation (N141I) linked to familial Alzheimer disease (Volga German families) increases the secretion of amyloid β protein ending at the 42nd (or 43rd) residue. Proc Natl Acad Sci USA 94: 2025–2030

The APP and PS1/2 Mutations Linked to Early Onset Familial Alzheimer's Disease Increase the Extracellular Concentration of Aβ1-42(43)

S. G. Younkin*

Introduction

The amyloid β protein (Aβ) is an ~4 kD secreted protein that is derived from a set of large, alternatively spliced precursor proteins collectively referred to as the amyloid β protein precursor (βAPP). Secreted Aβ is readily detected in cerebrospinal fluid (CSF), plasma, and medium conditioned by cultured cells (Seubert et al. 1992; Shoji et al. 1992; Haass et al. 1992; Busciglio et al. 1993). Most secreted Aβ is Aβ1-40, but a small component (5–10%) is Aβ1-42 (Dovey et al. 1993; Vigo-Pelfrey et al. 1993; Suzuki et al. 1994). A large amount of amyloid β protein (Aβ) is deposited extracellularly in the senile plaques that are invariably observed in the brains of patients with all forms of Alzheimer's disease (AD). Aβ1-42 appears to be particularly important in AD because it forms insoluble amyloid fibrils more rapidly than Aβ1-40 *in vitro* (Hilbich et al. 1991; Burdick et al. 1992; Jarrett et al. 1993; Jarrett and Lansbury 1993) and is deposited early and selectively in senile plaques (Iwatsubo et al. 1995). Extracellular Aβ deposition could be 1) an essential early event in AD pathogenesis, 2) an "innocent" marker that is invariably associated with some other change that drives AD pathogenesis, or 3) an unimportant, end-stage consequence of AD pathology. To examine the importance of Aβ in AD, we have analyzed the effect of the amyloid β protein (*APP;* Goate et al. 1991; Mullan et al. 1992), presenilin 1 (*PS1;* Sherrington et al. 1995) and presenilin 2 (*PS2;* Levey-Lahad et al. 1995; Rogaev et al. 1995) mutations that are known to cause early onset familial AD (FAD) on extracellular Aβ concentration.

Swedish βAPP$_{K670N, M671L}$ mutation

In 1992, Mullan et al. identified a large Swedish family in which the AD phenotype cosegregates with a double mutation that converts the lysine-methionine at βAPP670,671 to asparagine-leucine. The age of disease onset in this family is 53±4 years (mean ± SD) with a range from 44 to 61 years. Clinical progression is indistinguishable from other forms of AD, and the clinical diagnosis has been confirmed by neuropathologic examination of the brain of a deceased mutation

---

* Mayo Clinic Jacksonville, 4500 San Pablo Rd. Jacksonville, FL 32224, USA

carrier. Previous analyses of this mutation have shown 1) that $\beta APP_{K670N,M671L}$ undergoes altered processing in transfected cultured cells, releasing 5–6 times more Aβ1-40 and Aβ1-42(43) than wild type βAPP (Citron et al. 1992; Cai et al. 1993; Suzuki et al. 1994), and 2) that fibroblasts from affected and pre-symptomatic individuals with the $\beta APP_{K670N,M671L}$ mutation secrete more 4 kD Aβ than fibroblasts from age-matched controls (Citron et al. 1994). Thus there is good evidence that this mutation causes AD by coordinately increasing secretion of Aβ1-40 and Aβ1-42(43), thereby increasing Aβ concentration in a way that fosters amyloid deposition.

We have developed sandwich ELISAs that specifically detect fmol amounts of Aβ1-40 or Aβ1-42 in plasma (Scheuner et al. 1996) and in medium conditioned by cultured cells (Suzuki et al. 1994). Since the $\beta APP_{K670N,M671L}$ family is quite large and well characterized, it provided an excellent opportunity to determine whether measurement of plasma Aβ is useful in identifying individuals who develop AD because of elevated Aβ concentration. Plasma samples from 43 individuals in this family were analyzed (Scheuner et al. 1996); 12 of these individuals carried the $\beta APP_{K670N,M671L}$ mutation and 31 were non-carriers. In each of the 12 carriers, the concentration of Aβ1-40 in plasma (Fig. 1) was substantially higher than in any of the 31 non-carriers. Plasma levels of Aβ1-40 were 511±124 pM (mean ± SD) and 178±29 pM (p<0.0001) in carriers and non-carriers, respectively. In the 12 carriers, Aβ1-40 ranged from 329–752 pM, and there was no significant difference between the seven pre-symptomatic carriers

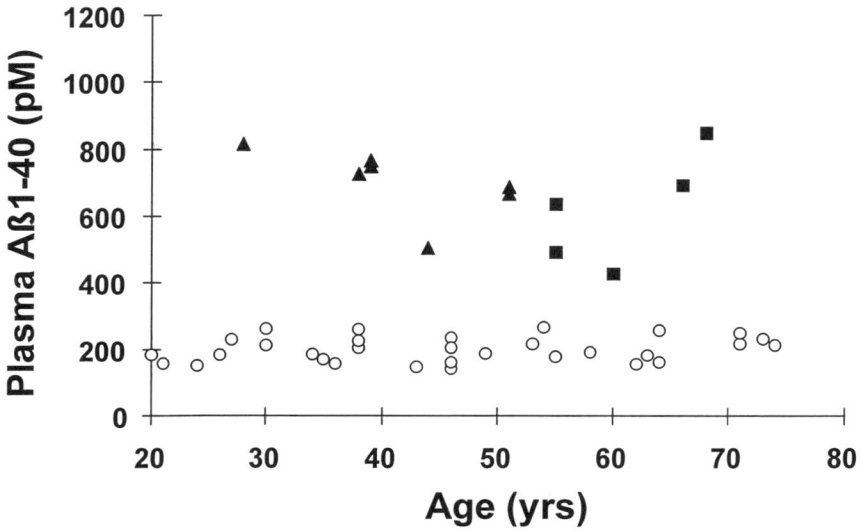

Fig. 1. Plasma Aβ1-40 concentration in the Swedish FAD ($\beta APP_{K670N,M671L}$) family. Non-carriers (O), pre-symptomatic carriers (▲), symptomatic carriers (■). Methods were as described by Scheuner et al. 1996

(522±82 pM), who were 28–51 years old, and the 5 symptomatic carriers (495±178 pM), who were 55–68 years old. In the 31 non-carriers, plasma Aβ1-40 ranged from 119–231 pM and it showed no obvious change with age, which ranged from 20–74 years.

The concentration of the Aβ1-42(43) in the plasma of non-carriers was 27±8 pM, it ranged from 18–59 pM, and it also showed no obvious change with age (Fig. 2). In the carriers, Aβ1-42 was significantly elevated at 57±16 pM ($p<0.0001$), and it ranged from 41–99 pM. Like Aβ1-40, Aβ1-42(43) was not significantly different in presymptomatic (52±5 pM) as compared to symptomatic carriers (64±24 pM).

To determine whether the apolipoprotein E4 (ApoE4) allele, which has been shown to influence the age of onset of AD (Pericak-Vance et al. 1991), influences the concentration of Aβ in plasma, we analyzed Aβ in 30 non-carriers (5 E4,E3; 3 E4,E2; 14 E3,E3; and 8 E3,E2), seven pre-symptomatic carriers (4 E4,E3; 2 E3, E3; and 1 E3,E2), and five symptomatic carriers (1 E4,E4; 2 E3,E3, and 2 E3,E2) whose ApoE genotype had been determined. We observed no significant effect of ApoE genotype on the concentration of Aβ1-40 or Aβ1-42(43) in plasma. This result is consistent with recent reports (Ma et al. 1994; Wisniewski et al. 1994; Castaño et al. 1995; Evans et al. 1995) that amyloid fibril formation occurs more rapidly *in vitro* in the presence of ApoE4 as compared to ApoE3, a finding that suggests that ApoE4 may act *in vivo* by influencing complexes that foster amyloid deposition rather than by altering processing in a way that increases Aβ concentration.

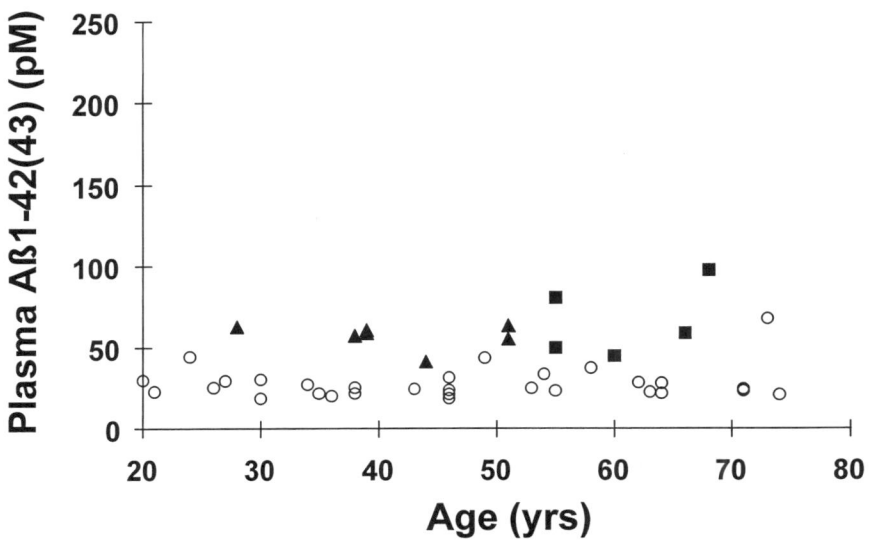

Fig. 2. Plasma Aβ1-42(43) concentration in the Swedish FAD (βAPP$_{K670N,M671L}$) family. Non-carriers (○), pre symptomatic carriers (▲), symptomatic carriers (■). Methods were as described by Scheuner et al. 1996

It has previously been reported that the concentration of Aβ in CSF declines as AD progresses (Motter et al. 1995). We therefore analyzed the carriers to determine if plasma Aβ1-40 or Aβ1-42(43) decrease in the Swedish (βAPP$_{K670N,M671L}$) family as cognitive performance, measured by Mini-Mental Status examination (MMSE), declines. Although no significant correlations were found, Aβ1-40 and to a lesser extent Aβ1-42(43) appeared to decline in the carriers as they approached the age of onset of AD and became slightly demented (MMSE >20).

Overall, this analysis of the Swedish FAD family shows that the βAPP$_{K670N,M671L}$ mutation produces a substantial, highly significant increase in plasma Aβ1-40 and Aβ1-42(43), and that this increase is present long before disease develops. By measuring plasma Aβ, one can, in fact, identify the carriers within the βAPP$_{K670N,M671L}$ family who are destined to develop AD. These findings, which are completely consistent with previous *in vitro* studies (Citron et al. 1992, 1994; Cai et al. 1993; Suzuki et al. 1994), show unequivocally that the βAPP$_{K670N,M671L}$ mutation increases Aβ concentration *in vivo*. Thus they provide strong additional evidence that βAPP$_{K670N,M671L}$ causes AD by increasing secretion of Aβ1-40 and Aβ1-42(43), thereby increasing Aβ concentration in a way that fosters amyloid deposition. Perhaps more importantly, the findings indicate that measurement of plasma Aβ is a powerful tool for identifying individuals who develop AD because Aβ concentration is elevated.

Other FAD-Linked Mutations

On this basis, we performed a second study (Scheuner et al. 1996) in which we analyzed nine subjects with one of four PS1 mutations, three subjects with the Volga German PS2$_{N141I}$ mutation, and one subject with a βAPP$_{V717I}$ mutation as compared with 14 controls. Plasma Aβ1-40 did not increase in subjects with PS1/2 or βAPPV717I mutations. The mean concentration of Aβ1-42(43) was significantly ($p<0.0001$) increased in the eight symptomatic subjects with PS1 mutations (PS1$_{G209V}$, PS1$_{M146V}$, PS1$_{H163R}$, or PS1$_{E120D}$). A similar increase in Aβ1-42(43) was observed in the presymptomatic subject with a PS1$_{G209V}$ mutation and the symptomatic subject with an βAPP$_{V717I}$ mutation. In the three subjects with PS2$_{N141I}$ mutations, mean plasma Aβ1-42(43) was also significantly ($p = 0.009$) increased.

Sporadic AD

To evaluate plasma Aβ in sporadic AD, we performed a third study (Scheuner et al. 1996) in which we analyzed plasma Aβ in 71 elderly patients with sporadic AD and 75 controls well matched for age, sex and ethnicity. We observed no significant difference in plasma Aβ1-40 or Aβ1-42(43) in the sporadic AD patients as compared to controls. Although most of the sporadic AD patients that we examined clearly did not have increased plasma Aβ42(43), inspection of the data

Table 1. Effect of ApoE genotype on plasma Aβ concentration

ApoE genotype	2/3	3/3	4/2	4/3	4/4
Aβ1–40	157 ± 7	196 ± 15	163 ± 10	166 ± 6	157 ± 8
Aβ1–42(43)	27 ± 5	30 ± 3	34 ± 7	25 ± 1	24 ± 1
No. Subjects	16	67	8	38	17
No. AD (%AD)	6 (38%)	21 (31%)	4 (50%)	23 (61%)	17 (100%)

from 71 sporadic AD patients and 75 controls showed that in 11 of the 146 subjects examined, Aβ1-42(43) was elevated into the range observed in subjects with the FAD-linked *APP, PS1*, and *PS2* mutations. In this group of 11, the frequency of sporadic AD was substantially and significantly ($p<0.03$) increased; nine of the 11 subjects had sporadic AD, the two unaffected individuals were younger subjects still at risk for AD, and the five subjects over the age of 80 all had sporadic AD. Remarkably, two of the nine subjects with elevated plasma Aβ1-42(43) showed this elevation before the onset of clinically apparent disease; they were in the control group initially and subsequently developed AD. Thus an elevated concentration of Aβ1-42(43) that is detectable in plasma may play an important role in 10–20% of sporadic AD cases, and this elevation may be present before symptoms develop.

To further evaluate the effect of the ApoE4 allele on plasma Aβ, we analyzed the sporadic AD subjects and the controls after categorizing them for their ApoE genotype (Table 1). In this series, as in our analysis of the Swedish FAD family, there was no indication that the ApoE4 allele increases Aβ1-40 or Aβ1-42(43). If anything, there was a slight reduction in Aβ1-40 and Aβ1-42(43) in the subjects with E4/E4 or E4/E3 genotypes as compared to subjects with E3/E3 genotypes (Table 1).

Discussion

Our results show unequivocally that a fundamental effect of the FAD-linked APP, PS1, and PS2 mutations is to increase the extracellular concentration of Aβ1-42(43) or of both Aβ1-40 and Aβ1-42(43) *in vivo*. This will foster the Aβ deposition that is invariably observed in all forms of AD. Thus our results provide strong evidence that Aβ, specifically Aβ42, plays an important early role in AD pathogenesis, making it very unlikely that Aβ deposition is an unimportant end-stage consequence of AD pathology. Although it is formally possible that the increases we have observed are a good "marker" generated by another process that produces AD independently of alterations in Aβ metabolism, this seems highly unlikely. Other observations that provide strong support for the concept that Aβ deposition in AD is likely to be toxic rather than an innocent epiphenomenon include 1) the finding that aggregated synthetic Aβ is toxic to cultured neurons *in vitro* and *in vivo*, 2) the observation that Aβ can trigger the classic complement cascade *in vitro*, and 3) the finding that the Aβ deposited in neuritic plaques is intimately associated with proteins of the classic complement cascade and with reactive microglia likely to be releasing cytokines and reactive free radicals that could have neurotoxic effects.

It is highly unlikely that cerebral Aβ42(43) deposition is a direct result of the increased plasma Aβ42(43) produced by the FAD-linked *APP* and *PS1/2* mutations. Instead, this deposition is almost certainly due to an increase in the extracellular concentration of Aβ42(43) in the brain that occurs as part of a generalized, system-wide effect of these mutations. This is supported by recent studies showing increased brain Aβ42(43) in transgenic mice expressing mutant as compared to wild type PS1 (Duff et al. 1996; Borchelt et al. 1996; Citron et al. 1997) and by studies showing increased secreted Aβ42(43) in a number of different transfected cell lines expressing mutant as compared to wild type PS1 (Borchelt et al. 1996; Citron et al. 1996).

References

Borchelt DR, Thinakaran G, Eckman CB, Lee MK, Davenport F, Ratovitsky T, Prada C-M, Kim G, Seekins S, Yager D, Slunt HH, Wang R, Seeger M, Levey AI, Gandy SE, Copeland NG, Jenkins NA, Price DL, Younkin SG, Sisodia SS (1996) Familial Alzheimer's disease-linked presenilin 1 variants elevate Aβ1-42/1-40 ratio in vitro and in vivo. Neuron **17**: 1005-1013

Burdick D, Soreghan B, Kwon M, Kosmoski J, Knauer M, Henschen A, Yates J, Cotman C, Glabe C (1992) Assembly and aggregation properties of synthetic Alzheimer's A4/β amyloid peptide analogs. J Biol Chem **267**: 546-554

Busciglio J, Gabuzda DH, Matsudaira P, Yankner BA (1993) Generation of β-amyloid in the secretory pathway in neuronal and nonneural cells. Proc Natl Acad Sci USA **90**: 2092-2096

Cai XD, Golde TE, Younkin SG (1993) Release of excess amyloid β protein from a mutant amyloid beta protein precursor. **Science 259**: 514-516

Castaño E, Prelli F, Wisniewski T, Golabek A, Kumar RA, Soto C, Frangione B (1995) Fibrillogenesis in Alzheimer's disease of the amyloid β peptides and apolipoprotein E. **Biochem J 306**: 599-604

Citron M, Oltersdorf T, Haass C, McConlogue L, Hung AY, Seubert P, Vigo-Pelfrey C, Lieberburg I, Selkoe DJ (1992) Mutation of the beta-amyloid precursor protein in familial Alzheimer's disease increases β-protein production. **Nature 360**: 672-674

Citron M, Vigo-Pelfrey C, Teplow DB, Miller C, Schenk D, Johnston J, Winblad B, Venizelos N, Lannfelt L, Selkoe DJ (1994) Excessive production of amyloid beta-protein by peripheral cells of symptomatic and presymptomatic patients carrying the Swedish familial Alzheimer disease mutation. **Proc Natl Acad Sci USA 91**: 11993-11997

Citron M, Westaway D, Xia W, Carlson G, Diehl T, Levesque G, Johnson-Wood K, Lee M, Seubert P, Davis A, Kholodenko D, Motter R, Sherrington R, Perry B, Yao H, Strome R, Lieberburg I, Rommens J, Kim S, Schenk D, Fraser P, St George Hyslop P, Selkoe DJ (1997) Mutant presenilins of Alzheimer's disease increase production of 42-residue amyloid β-protein in both transfected cells and transgenic mice. **Nature Med 3**: 67-72

Dovey HF, Suomesaari-Chrysler S, Lieberburg I, Hinha S, Kiem PS (1993) Cells with a familial Alzheimer's disease mutation produce authentic β-peptide. **NeuroReport 4**: 1039-1042

Duff K, Eckman C, Zehr C, Yu X, Prada C-M, Perez-Tur J, Hutton M, Buee L, Harigaya Y, Yager D, Morgan D, Gordon MN, Holcomb L, Refolo L, Zenk B, Hardy J, Younkin S (1996) Increased amyloid-β42(43) in brains of mice expressing mutant presenilin 1. **Nature 383**: 710-713

Evans KC, Berger EP, Cho CG, Weisgraber KH, Lansbury Jr, PTL (1995) Apolipoprotein E is a kinetic but not a thermodynamic inhibitor of amyloid formation: implications for the pathogenesis and treatment of Alzheimer's disease. **Proc Natl Acad Sci USA 92**: 763-767

Goate A, Chartier-Harlin MC, Mullan M, Brown J, Crawford F, Fidani L, Giuffra L, Haynes A, Irving N, James L, Mant R, Newton P, Rooke K, Roques P, Talbot C, Pericak-Vance M, Roses A, Williamson R, Rossor M, Owen M, Hardy J (1991) Segregation of a missense mutation in the amyloid precursor protein gene with familial Alzheimer's disease. **Nature 349**: 704–706

Haass C, Schlossmacher MG, Hung AY, Vigo-Pelfrey C, Mellon A, Ostraszewski BL, Lieberburg I, Koo EH, Schenk D, Teplow DB, Selkoe DJ (1992) Amyloid β-peptide is produced by cultured cells during normal metabolism. **Nature 359**: 322–325

Hilbich C, Kisters-Woike B, Reed J, Masters CL, Beyreuther K (1991) Agggregation and secondary structure of synthetic amyloid βA4 peptides of Alzheimer's disease. **J Mol Biol 218**: 149–163

Iwatsubo T, Mann DM, Odaka A, Suzuki N, Ihara Y (1995) Amyloid β protein (A β) deposition: A β 42(43) precedes A β 40 in Down syndrome. **Ann Neurol 37**: 294–299

Jarrett JT, Lansbury PT Jr (1993) Seeding "one dimensional crystallization" of amyloid: a pathogenic mechanism in Alzheimer's disease and Scrapie? **Cell 73**: 1055–1058

Jarrett JT, Berger EP, Lansbury PT Jr (1993) The carboxy terminus of β amyloid protein is critical for the seeding of amyloid formation: implications for pathogenesis of Alzheimer's disease. **Biochemistry 32**: 4693–4697

Levy-Lahad E, Wijsman EM, Nemens E, Anderson L, Goddard KA, Weber IL, Bird TD, Schellenberg GD (1995) A familial Alzheimer's disease locus on chromosome 1. **Science 269**: 970–973

Ma J, Yee A, Brewer HB Jr., Das S, Potter H (1994) Amyloid-associated proteins α 1-antichymotrypsin and apolipoprotein E promote assembly of Alzheimer β-protein into filaments. **Nature 372**: 92–94

Motter R, Vigo-Pelfrey C, Kholodenko D, Barbour R, Johnson-Wood K, Galasko D, Chang L, Miller B, Clark C, Green R et al. (1995) Reduction of β-amyloid peptide 42 in the cerebrospinal fluid of patients with Alzheimer's disease. **Ann Neurol 38**: 643–648

Mullan M, Crawford F, Axelman K, Houlden H, Lilius L, Winblad B, Lannfelt L (1992) A pathogenic mutation for probable Alzheimer's disease in the APP gene at the N-terminus of β-amyloid. **Nature Genet 1**: 345–347

Pericak-Vance MA, Bebout JL, Gaskell PC, Yamaoka LH, Hung W-Y, Alberts MJ, Walker AP, Bartlett RJ, Haynes CA, Welst KA, Earl NL, Heymark A, Clark CM, Roses AD (1991) Linkage studies in familial Alzheimer's disease: evidence for chromosome 19 linkage. Am J Hum Genet 48: 1034–1050

Rogaev E, Sherrington R, Rogaeva EA, Levesque G, Ikeda M, Liang Y, Chi H, Lin C, Holman K, Tsuda T, Mar L, Sorbi S, Nacmias B, Piacentini S, Amaducci L, Chumakov I, Cohen D, Lannfelt L, Fraser PE, Rommens JM, St George-Hyslop PH (1995) Familial Alzheimer's disease in kindreds with missense mutations in a gene on chromosome 1 related to the Alzheimer's disease type 3 gene. **Nature 376**: 775–778

Seubert P, Vigo-Pelfrey C, Esch F, Lee M, Dovey H, Davis D, Sinha S, Schlossmacher M, Whaley J, Swindlehurst C, McCormack R, Wolfert R, Selkoe D, Lieberburg I, Schenk D (1992) Isolation and quantification of soluble Alzheimer's β-peptide from biological fluids. **Nature 359**: 325–327

Sherrington R, Rogaev EI, Liang Y, Rogaeva EA, Levesque G, Ikeda M, Chi H, Lin C, Li G, Holman K, Tsuda T, Mar L, Foncin J-F, Bruni AC, Montesi MP, Sorbi S, Rainero I, Pinessi L, Nee L, Chumakov Y, Pollen D, Wasco W, Haines JL, Da Silva R, Pericak-Vance M, Tanzi RE, Roses AD, Fraser PE, Rommens JM, St George-Hyslop PH (1995) Cloning of a novel gene bearing missense mutations in early onset familial Alzheimer disease. **Nature 375**: 754–760

Shoji M, Golde TE, Ghiso J, Cheung TT, Estus S, Shaffer LM, Cai XD, McKay DM, Tintner R, Frangione B, Younkin SG (1992) Production of the Alzheimer amyloid beta protein by normal proteolytic processing. **Science 258**: 126–129

Suzuki N, Cheung TT, Cai XD, Odaka A, Otvos L Jr., Eckman C, Golde TE, Younkin SG (1994) An increased percentage of long amyloid β protein secreted by familial amyloid β protein precursor (β APP717) mutants. **Science 264**: 1336–1340

Vigo-Pelfrey C, Lee D, Keim P, Lieberburg I, Schenk DB (1993) Characterization of β amyloid peptide from human cerebrospinal fluid. **J Neurochem 61**: 19965–19968

Wisniewski T, Castano EM, Golabek A, Vogel T, Frangione B (1994) Acceleration of Alzheimer's fibril formation by apolipoprotein E in vitro. **Am J Pathol 145**: 1030–1035

Metabolism and Function of Presenilin 1

S. S. Sisodia, G. Thinakaran, P. C. Wong, D. R. Borchelt, M. K. Lee, A. Doan, J. Regard, H. Chen, H. Zheng, C. Eckman, H. H. Slunt, T. Ratovitsky, F. Davenport, C. Harris, L. H. T. Van der Ploeg, S. G. Younkin, N. A. Jenkins, N. G. Copeland, and D. L. Price[*]

Summary

Neither the normal functions of presenilins nor the mechanism(s) by which familial Alzheimer's disease (FAD)-linked mutations cause AD have been defined. Presenilin 1 (PS1) is a polytopic membrane protein that is subject to endoproteolytic processing *in vivo;* PS1 derivatives accumulate to saturable levels and to ~ 1:1 stoichiometry by mechanism(s) that are not fully defined. We show here that the two fragments coassemble. Moreover, we have detected neither interactions between PS1/PS2 and amyloid precursor protein (APP) nor influences of presenilin expression on APP maturation/secretion. To examine the *in vivo* function(s) of PS1, we developed mice with functionally inactivated *PS1* alleles. These animals die before birth and exhibit several developmental defects, including a poorly differentiated vertebral column, a phenotype traced to abnormal segmentation of somites. Whole mount *in situ* hybridization analyses reveal that specification of somitic cell lineages is apparently unaffected, despite the clear disruption in somite segmentation. However, notable differences in expression of *Notch1* and *Dll1* mRNAs were observed in $PS1^{-/-}$ embryos; in contrast to wild-type embryos in which abundant expression of *Notch1* and *Dll1* mRNAs are observed in the presomitic mesoderm, the expression of these genes is nearly abolished in the $PS1^{-/-}$ embryos. Hence, PS1 serves to regulate the spatiotemporal expression of *Notch1* and *Dll1* in the paraxial mesoderm. Finally, we failed to detect any differences in the levels of $A\beta42$ and $A\beta40$ in brains of mice heterozygous for *PS1* relative to wild-type littermates. Thus, mutations in PS1 probably cause AD not by the loss but rather by the gain of deleterious function of mutant polypeptides.

---

[*] From the Departments of Pathology (SSS, GT, PCW, DRB, MKL, DLP), Neurology (DLP), Neuroscience (SSS, JR, DLP), and the Neuropathology Laboratory (SSS, GT, PCW, DRB, MKL, AD, HC, HHS, TR, FD, CH, DLP), The Johns Hopkins University School of Medicine, Baltimore, Maryland; Merck Research Labs (HZ, LHTV), Rahway, New Jersey; Mayo Clinic (CE, SGY), Jacksonville, Florida; Mammalian Genetics Laboratory, ABL-Basic Research Program, NCI-Frederick Cancer Center Research & Development (NAJ, NGC), Frederick, Maryland

Introduction

Alzheimer's disease (AD), the most common cause of dementia in the elderly, is associated with several risk factors, including age and inheritance. The majority of early-onset cases of AD (onset <60 years) are inherited as autosomal dominant disorders. To date, mutations have been identified in three genes that cosegregate with affected members of FAD pedigrees: the *APP* gene on chromosome 21 (Goate et al. 1991; Chartier-Harlin et al. 1991; Naruse et al. 1991; Mullan et al. 1992; Hendricks et al. 1992), the *PS1* gene on chromosome 14 (Schwab 1977; Sherrington et al. 1995), and the *PS2* gene on chromosome 1 (Levy-Lahad et al. 1995; Rogaev et al. 1995). Although mutations in *APP* cosegregate with ~ 19 pedigrees with FAD, mutations in *PS1* are causative in ~ 25-30% of pedigrees with early-onset FAD (Schellenberg 1995). Over 35 missense mutations (Campion et al. 1995; Chapman et al. 1995; Clark et al. 1995; Cruts et al. 1995; Sherrington et al. 1995; Wasco et al. 1995; Boteva et al. 1996) and a point mutation upstream of a splice acceptor site that results in an inframe deletion of exon 9 (PS1ΔE9; Perez-Tur et al. 1995) have been identified in the *PS1* gene in families with early onset FAD.

The normal function(s) of presenilins in vertebrates has not been defined. In this regard, a homologous gene in *C. elegans*, termed *sel-12*, has been identified that facilitates signalling mediated by the lin-12/Notch family of receptors involved in developmental cell fate specification and lateral inhibition (Levitan and Greenwald 1995). Moreover, the mechanism(s) by which FAD-linked mutations cause AD is unresolved. The absence of nonsense or frameshift mutations leading to truncated PS1/PS2 supports the notion that AD is caused not by the loss but rather by the gain of deleterious function of mutant polypeptides. In support of this view, studies of Aβ_{40} and Aβ_{42} (43) production in the conditioned medium from fibroblasts or plasma of affected members of pedigrees with *PS1/PS2*-linked mutations (Scheuner et al. 1996) transfected mammalian cells, and the brains of transgenic mice (Borchelt et al. 1996; Duff et al. 1996; Citron et al. 1997; Tomita et al. 1997) reveal that one mechanism by which mutant PS1 cause AD is by the acquisition (or enhancement) of property(ies) that influence APP processing in a manner that leads to increased extracellular concentrations of Aβ_{42} (43). In addition, ICC studies have demonstrated massive Aβ_{42} (43) deposition in the cerebral cortex and severe cerebellar pathology, including Aβ_{42}-reactive plaques, many bearing dystrophic neurites and reactive glia in individuals with a PS1-linked E280A mutation (Lemere et al. 1996). These data suggest that the FAD-linked PS1/PS2 variants influence processing at the "γ-secretase" site and cause AD by increasing the extracellular concentration of highly amyloidogenic Aβ_{42} (43) species, thus fostering Aβ amyloid deposition in the brain.

Presenilin Structure

Secondary structure algorithms predict that PS1 contains between seven (Sherrington et al. 1995) and nine (Slunt et al. 1995) transmembrane domains, including a hydrophilic acidic "loop" region encompassing amino acids 263–407. We examined the topology of PS1 using two strategies: first, putative "transmembrane helices" were tested for their ability to export a protease-sensitive substrate across a lipid bilayer; and second, the plasma membrane of cultured cells expressing human PS1 or the FAD-linked PS1ΔE9 variant were selectively permeabilized, and the accessibility of antibodies specific for the N-terminus "loop" and C-terminus to cognate epitopes was analyzed by indirect immunofluorescence microscopy. These studies established that the N-terminus, "loop," and C-terminus of PS1 are cytoplasmic (Doan et al. 1996). In parallel, the topology of the *C. elegans* presenilin homolog, termed "sel-12," was determined using a series of sel-12 β-galactosidase chimeras, an approach that relies on the observation that β-galactosidase is active in the cytoplasm of cells and is inactive in the lumen of membrane compartments. The deduced topology of sel-12 indicated that the protein spans the membrane eight times, with the N- and C-termini being exposed to the cytosol (Li and Greenwald, 1996). Based on the high homology of the first six transmembrane domains between PS1, PS2, and sel-12, these regions may mediate the tertiary structure important for functional activity. A unique feature of the topology model is the predicted association of a potential transmembrane helix near the N-terminus of the "loop" domain with the lipid bilayer. This type of association of a putative transmembrane helix is highly reminiscent of putative transmembrane helix (TMDII) of the subunits of AMPA glutamate receptors, GluR1 (Hollmann et al. 1994) and the goldfish kainate binding proteins (GFKARα and GVKARβ) (Wo and Oswald, 1994), and has recently proven for the ionotropic NMDA receptor channel M2 segment (Kuner et al. 1996).

Endoproteolysis of PS1

The biosynthesis and metabolism of PS1 has been investigated in cultured cells and *in vivo*. Surprisingly, although PS1 is synthesized as an ~ 42–43 kD polypeptide, the preponderant PS1-related species that accumulate in cultured mammalian cells and in brain and systemic tissues of rodents, primates, humans, and transgenic mice are ~ 27–28 kD N-terminal (NTF) and ~ 16–17 kD C-terminal (CTF) derivatives (Duff et al. 1996; Lee et al. 1996; Mercken et al. 1996; Thinakaran et al. 1996; Citron et al. 1997; Hendricks et al. 1997; Lah et al. 1997; Tomita et al. 1997; Levey et al. 1997). Epitope mapping studies suggest that PS1 is cleaved within a region encompassing amino acids 260–320, a domain in which over 50% of identified FAD-linked *PS1* mutations occur. Studies suggest that cleavage occurs between amino acids 298 and 299 (D. J. Selkoe, personal communication). These results are consistent with the demonstration that the FAD-linked PS1ΔE9 variant that lacks amino acids 290–319 fails to be cleaved (Thinakaran et al. 1996). Notably, the accumulation of ~17 kD and ~27 kD human-specific PS1 derivatives in brains of transgenic mice expressing human PS1 is highly regulated and saturable; the levels of PS1 derivatives are disproportionate to the levels of transgene-derived mRNA of full-length human PS1. Moreover, the stoichiometry of accumulated ~17 kD and ~27 kD PS1 fragments is ~1:1 in nontransgenic and transgenic mouse brain, and this ratio is independent of the level of transgene-derived human PS1 mRNA expression (Thinakaran et al. 1996). The mechanism(s) involved in regulating the levels of accumulated PS1-derivatives has not been established.

A conserved feature of the topology models for the presenilin homologues is that the site for endoproteolytic cleavage is located in the cytosolic compartment. At present, neither the identity of the protease nor the physiological significance of PS1 proteolysis is known. However, and in view of the demonstration of a paucity of full length PS1 and highly regulated accumulation of processed derivatives *in vivo*, it is highly likely that the PS1 fragments are the functional units (Thinakaran et al. 1996). Moreover, evidence that human PS2 (Citron et al. 1997; Tomita et al. 1997; Kim et al. 1997) and sel-12-β-galactosidase chimeras (Li and Greenwald, 1996) are also subject to endoproteolytic cleavage indicate that endoproteolysis is a highly conserved process and arguably, a processing event that regulates PS accumulation (see below).

Association of PS1 NTF and CTF

Our demonstration that the accumulation of PS1 NTF and CTF is coregulated to 1:1 stoichiometry, independent of the level of full-length PS1, suggested that the two derivatives may be coresident *in vivo*. We utilized two strategies to address this issue. First, we treated untransfected African green monkey kidney, COS-1, cells with a membrane permeable, sulfhydryl-cleavable crosslinking agent, DSP, then lysed cells with an SDS-containing detergent cocktail. PS1-related polypeptides were immunoprecipitated from the lysate with an N-terminal antibody, Ab14, and the resulting immunocomplex was treated with β-mercaptoethanol and then fractionated by SDS-PAGE. Fractionated proteins were subject to immunoblotting with αPS1Loop, an antibody specific for epitopes between amino acids 320–380 of the PS1 loop domain. We documented that the PS1 CTF was co-isolated with the PS1 NTF using this paradigm. Conversely, we documented that the PS1 NTF could be coprecipitated along with the PS1 CTF after crosslinking and detergent solubilization. Concerned with the specificity of the crosslinking procedure, we probed blots containing immunoisolated proteins with antibodies specific for APP. We failed to detect any endogenous monkey APP-immunoreactive species in this experiment, despite the demonstration that APP is easily detectable following direct immunoblot analysis of one-tenth of the lysate used for the coimmunoprecipitation analysis.

Second, we employed a strategy to demonstrate the association of PS1 NTF and CTF in the presence of mild detergents. In this case, the PS1 NTF and CTF were immunoprecipitated from lysates of mouse neuroblastoma cells, N2a, that coexpress human PS1 and human APP. Cells were lysed in 0.25% N-dodecyl 2-D maltoside, in the absence of crosslinkers or stabilizers. Under these conditions, we clearly demonstrated that the PS1 NTF and CTF could be coisolated. In parallel investigations, we have confirmed that the PS2 NTF and CTF can be coisolated from 0.25% N-dodecyl maltoside lysates of N2a cells expressing human PS2. Notably, we failed to document the coisolation of human APP, an integral membrane protein, or superoxide dismutase 1 (SOD1), an abundant cytosolic protein from human PS1- or human PS2-expressing N2a cells.

In support of the coimmunoprecipitation experiments in cultured cells, gel filtration chromatography and sucrose density gradient fractionation of PS1 isolated from cells extracted in mild detergents or homogenized in isotonic buffers, respectively, reveal that the PS1 N- and C-terminal fragments appear to remain associated in larger complexes of ~ 100 kD (Seeger et al. 1997). It is not presently clear whether the coisolated fragments are in homomeric or heteromeric assemblies.

Mechanisms of PS1 NTF and CTF Stabilization

We examined the fate of mouse PS1 and PS2 in mouse neuroblastoma cell lines that overexpress human PS1 or PS2. Remarkably, murine PS1 and PS2 derivatives fail to accumulate in these cells. These observations in cultured cells have been confirmed by the demonstration that murine PS1 and PS2 are "replaced" by human PS1 derivatives in brains of transgenic mice expressing human PS1.

An important issue in the use of models of cultured cells that overexpress membrane-bound or secretory proteins is the potential artifact of overloading the endoplasmic reticulum (ER) and/or other compartments of the secretory pathway. Since PS1 and PS2 largely reside in the ER, it is conceivable that overexpression of either polypeptide could severely compromise the synthesis or steady-state levels of endogenous PS1, PS2, and/or other ER-resident polypeptides. To document the specificity of PS1/PS2 overexpression on murine PS1/PS2 accumulation, we performed several control experiments. We now document that neither the steady-state levels of PS1/PS2 mRNA, overall protein synthesis, nor the steady-state levels of membrane-bound or lumenal ER resident proteins, i.e., calnexin and Grp78/Bip, respectively, are altered in these cell lines.

One potential mechanism by which overexpression of human PS1 or PS2 might lead to the loss of endogenous PS1 or PS2 derivatives and concurrent "replacement" with human PS1 derivatives is by competing with limiting levels of the protease responsible for endoproteolytic cleavage. To address this issue, we generated stable N2a lines harboring a modified PS1 cDNA encoding a C-terminally truncated PS1 polypeptide ($PS1_{360}\Delta$) in which the last 107 amino acids of human PS1 (including part of the large hydrophilic "loop" domain, the last two transmembrane domains and the C-terminus) were deleted. The rationale for this experiment was to ask whether a truncated PS1 that contained a native site for endoproteolysis could effectively "compete" with endogenous full-length murine PS1/PS2 for endoproteolysis, hence diminishing the levels of murine PS1/PS2 derivatives. Although the PS1$_{360}\Delta$ polypeptide contains the site of cleavage (between residues 298 and 299) and extends 60 amino acids C-terminus to the cleavage site, immunoblot analysis using a monoclonal antibody specific for human PS1 N-terminal region failed to reveal the presence of ~ 28 kDa human PS1 NTF. Thus, PS1$_{360}\Delta$ molecules are not subject to endoproteolytic cleavage. Furthermore, and despite high levels of accumulation of PS1$_{360}\Delta$ molecules, the steady-state levels of murine PS1 derivatives remained unchanged. Essentially identical results were obtained by analysis of stable lines expressing PS1$_{403}\Delta$, a truncated PS1 polypeptide lacking the C-terminal 64 amino acids of human PS1. These studies raised the suggestion that endoproteolysis is a prerequisite for replacement of murine PS1 derivatives. However, we showed that, in stable N2a cell lines expressing an FAD-linked PS1 variant (PS1ΔE9) that fails to undergo endoproteolysis, the accumulation of endogenous PS1 derivatives was compromised. Thus, endoproteolysis is not entirely a prerequisite for "replacement."

Although little information is available regarding the molecular identity of the enzyme(s) responsible for PS1/PS2 endoproteolysis and factors responsible for fragment stability, we consider less likely the view that the regulated accumulation of PS fragments is dependent on competition for protease(s) responsible for cleavage. We demonstrate that expression of C-terminally truncated human PS1 polypeptides (PS1$_{360}\Delta$ and PS1$_{403}\Delta$) that contain the proteolytic cleavage site but which fail to be cleaved does not influence murine PS1/PS2 fragment accumulation. On the other hand, expression of a non-cleavable FAD-linked PS1ΔE9 variant results in the replacement of murine PS1/PS2 derivatives, and offers an attractive model wherein PS1ΔE9 polypeptides are incorporated to similar, if not identical, cellular compartments where PS1 derivatives are generated, and subsequently accumulate. Under these circumstances, the PS1ΔE9 molecule effectively outcompetes the newly-synthesized full-length murine PS1/PS2 molecules for limiting factors responsible for stabilization. In this setting, the PS1ΔE9 molecule behaves as the functional unit. On the other hand, C-terminally truncated (PS1$_{360}\Delta$ and PS1$_{403}\Delta$) molecules never "arrive" at the cleavage/stabilization compartment, because they are either misfolded or lack sequences that are required for trafficking to the appropriate compartments. In support of this view, we and others have demonstrated that although the PS1ΔE9 variant efficiently rescues an egg-laying defect in *C. elegans* lacking SEL-12 (Levitan et al. 1996; Baumeister et al. 1997), the human PS1 lacking the "loop" and C-terminal domains fails to do so (Baumeister et al. 1997). Thus, it is highly likely that the "functional" rescue by the PS1ΔE9 variant is a reflection of its stabilization in a limiting cellular compartment that is normally occupied by endoproteolytic derivatives of PS.

Our studies offer several attractive, though not mutually exclusive, hypotheses regarding the mechanisms underlying the cross-regulation of PS1/PS2 fragment accumulation, including: full-length PS1 and PS2 compete for a limiting cellular pool of factors necessary for endoproteolytic processing; and stabilization of both PS1 and PS2 derivatives is regulated by the association of shared, but limiting, cellular factors. Characterization of the protein trafficking signals and the enzyme(s) responsible for endoproteolysis of PS1 and PS2 will facilitate future efforts aimed at clarifying the limiting step(s) that regulates the accumulated levels of processed PS1/PS2 derivatives.

PS1 Function

The first major insight regarding presenilin function emerged from the discovery of the *C. elegans sel-12* gene that was uncovered in a suppressor screen aimed at reversing a lin-12 hypermorphic phenotype (Levitan and Greenwald 1995). Lin-12 is highly homologous to Notch, a member of a family of transmembrane receptors required for specification of cell fate and lateral inhibition during development (Artavanis-Tsakonas et al. 1995). Although details regarding the molecular mechanisms by which sel-*12* facilitates signalling mediated by lin-12 have not been established, it is conceivable that sel-12 could regulate lin-12 trafficking and cell-surface expression or, alternatively, act in a signalling capacity to modulate pathways activated by the binding of lin-12 to cognate ligands. The extremely high amino acid homology between the presenilins and sel-12, particularly in the first six transmembrane domains and the C-terminal $\sim$ 90 residues, led to the prediction that the related proteins would be functionally interchangeable. Consistent with this hypothesis, an egg-laying defect associated with loss of sel-12 function in *C. elegans* is efficiently rescued by the expression of human PS1 and PS2; the rescue efficiency of wild-type human PS1 and PS2 is essentially indistinguishable to transgenic worms expressing sel-12 (Levitan et al. 1996; Baumeister et al. 1997); notably, the egg-laying defect was only weakly rescued in transgenic worms expressing several human FAD-linked PS1 variants (Levitan et al. 1996; Baumeister et al. 1997). The PS1ΔE9 variant was the exception, showing considerable rescue activity relative to other PS1 missense variants. This latter observation argues that endoproteolytic cleavage is not obligatory for PS1 function (Levitan et al. 1996; Baumeister et al. 1997).

In view of the structural and functional homology between sel-12 and the presenilins, a role for presenilins in mammalian development was anticipated. *In situ* hybridization studies and RT-PCR approaches in mouse embryos reveal that presenilins are expressed in a ubiquitous manner during embryonic development, and as early as E8.5 (Lee et al. 1996; Wong et al. 1997). Most notably, the general spatial and temporal expression patterns of presenilins mRNAs do not directly coincide with the expression patterns of any specific member of the known mammalian Notch homologs (Reaume et al. 1992; Weinmaster et al. 1992; Lardelli et al. 1994; Lindsell et al. 1995; Williams et al. 1995). For example, although Notch1 mRNA expression in the developing neural tube is limited to cells near the ventricular zone, PS1 mRNAs are expressed in neuroblasts in the ventricular zones as well as in terminally differentiated neurons. These results suggest that PS1 function is not limited to Notch signalling alone. Moreover, since PS1 mRNA is expressed widely, it is unlikely that PS1 is restricted to signalling mediated by a single member of the Notch family, and it is more reasonable to suggest that PS may serve to mediate signalling by multiple Notch family members.

To examine the *in vivo* role of PS1 in mammalian development, we generated mice with a targeted disruption of the *PS1* gene ($PS1^{-/-}$ mice; Wong et al. 1997). Although homozygous mutant mice failed to survive beyond day one after birth, $PS1^{-/-}$ embryos were present at expected Mendelian frequencies at various stages of gestation ranging from E8.5 to E18.5; developmental deficits have not been observed in mice with a hemizygous mutation of *PS1*. The most striking phenotype observed in $PS1^{-/-}$ embryos was a severe perturbation in the development of the axial skeleton and ribs. The axial skeleton is drastically shortened and poorly segmented, while the ribs were poorly ossified, bifurcated, and malformed. The failed development of the axial skeleton in $PS1^{-/-}$ animals was traced to defects in the differentiation and condensation of the sclerotome. The delayed condensation of sclerotome and the observation of fused DRGs in $PS1^{-/-}$ embryos suggested that these phenotypes might have arisen as a consequence of earlier deficits in somitogenesis. This concept was confirmed by the observation that, in E8.5 and E9.5 embryos, somites were irregularly shaped and misaligned along the entire length of the neural tube; somites were largely absent at the caudalmost regions. The abnormal somite patterns in $PS1^{-/-}$ embryos were highly reminiscent of a phenotype described in mice with functionally inactivated *Notch1* alleles in which a delay and lack of coordination in somite segmentation was apparent (Conlon et al. 1995).

Whole mount *in situ* hybridization studies revealed that the expression of genes that identify specific somitic lineages (i.e., *PAX1* for sclerotome, *myogenin* for myotome) was unaffected in $PS1^{-/-}$ mice. On the other hand, expression of mRNA encoding Notch1 and Dll1, a Notch ligand, were dramatically reduced in the presomitic mesoderm of $PS1^{-/-}$ mice. The presomitic mesoderm was a loosely packed population of mesenchymal cells that differentiate into somites, a highly ordered epithelial structure. Thus, the loss of PS1 expression in $PS1^{-/-}$ mice affected the spatial and temporal expression of Notch1 and Dll1 in the presomitic mesoderm.

In addition to the defects in the axial skeleton, post E11.5 $PS1^{-/-}$ embryos exhibited hemorrhages limited to the brain and/or spinal cord; these lesions were not seen in $PS1^{+/-}$ or $PS1^{+/+}$ littermates. The underlying cellular and molecular defects responsible for the CNS hemorrhages in $PS1^{-/-}$ mice have not been determined. Nonetheless, this phenotype has been consistently observed in mice in which the genes encoding mammalian ligands of *Notch* have been ablated by gene targeting strategies. The murine ligands of *Notch*, termed *Jagged* and *Dll1*, are homologs of *Drosophila Serrate* and *Delta*, respectively. *Jagged*–/– embryos die at ~ E9.5 and exhibit hemorrhages limited to the CNS, while *Dll1*–/– embryos showed defects in somitogenesis, CNS hemorrhage and die at ~ E11.5

(T. Gridley and A. Gossler, personal communication). The similarities in phenotypes of $PS1^{-/-}$ and $Notch1^{-/-}$ mice, and $PS1^{-/-}$ and $Jagged^{-/-}$ or $Dll1^{-/-}$ alleles, suggest that PS1 may play a role in regulating the spatial and temporal expression of *Notch1* and *Dll1/Jagged* in the presomitic mesoderm and in the development of the CNS vasculature, respectively.

Presenilins in the Pathogenesis of AD

The mechanisms by which FAD-linked PS1 and PS2 variants cause AD are unclear. However, studies of Aβ40 and Aβ42 (43) production in the conditioned medium from fibroblasts of plasma of affected members of pedigrees with *PS1/PS2*-linked mutations (Scheuner et al. 1996), from transfected mammalian cells and from the brains of transgenic mice expressing FAD-linked presenilin variants (Borchelt et al. 1996; Duff et al. 1996; Citron et al. 1997; Tomita et al. 1997) reveal that one mechanism by which mutant PS1 cause AD is by the acquisition (or enhancement) of property(ies) that influence APP processing in a manner that leads to increased extracellular concentrations of highly fibrillogenic Aβ42 (43) peptides.

Our demonstration that mutant presenilins exhibited reduced (but not fully disrupted) rescue of an egg-laying defect in *C. elegans* lacking sel-12 suggested that FAD-linked presenilin variants may cause AD by a loss of function mechanism. To address this issue directly, we bred transgenic mice that express human APP to mice with a heterozygous disruption of endogenous *PS1*. The rationale for this experiment is based solely on the hypothetical model that PS1 functions to regulate (or limit) the production of Aβ42/43; this haploinsufficiency model predicts that loss of a functional allele might lead to increased Aβ1–42 production. Using a sensitive sandwich ELISA, we have failed to detect any differences in the absolute levels of Aβ42 and Aβ40 in brains of mice heterozygous for *PS1* relative to wild-type littermates. These studies lead us to conclude that mutations in PS1 cause AD not by loss but rather by gain of deleterious function of mutant polypeptides. In support of this view, nonsense or frameshift mutations leading to truncated PS1/PS2 have not been identified in FAD pedigrees.

Conclusions

Despite the wealth of information that has accrued over the last two years regarding the biology of presenilins, there are remarkable lacunae in our understanding of the molecular mechanisms that regulate presenilin biology in normal and disease settings. Several of these unresolved issues include: the mechanisms that regulate presenilin endoproteolysis and accumulation; the mechanisms by which PS1 influences Notch expression during development; the physiology of wild-type and mutant presenilin in the adult CNS; the identification and regulation of proteins that directly associate with PS1 and PS2; and the cell biological mechanisms by which mutant presenilins alter APP trafficking/metabolism to promote Aβ1-42 production. Clearly, the generation of transgenic mice expressing mutated PS1 and PS2 that recapitulate the neuropathological features of AD is an essential and extremely important goal. Biochemical, neuropathological and behavioral evaluation of these animals will provide important information regarding the pathogenic mechanisms of AD and serve as a foundation for the design of rational therapeutic strategies aimed at ameliorating this devastating neurodegenerative disorder of the elderly.

Acknowledgements. The authors thank Drs. Samuel E. Gandy (Cornell Medical College, New York, NY), Mary Seeger (Cornell University Medical College, New York, NY), Allan Levey (Emory University School of Medicine, Atlanta, GA), Christian Haass (University of Heidelberg, Mannheim, Germany), and Steven Wagner (SIBIA, La Jolla, CA) for antibodies used in these studies.
This work was supported by grants from the U.S. Public Health Service, National Institute of Health grants NIH AG 14248, AG 05146, NS 20471, P01 AG 14633-01 AG 12685-04, and AG 06656, and by grants from the Adler Foundation, the Devilbiss Fund, the Alzheimer's Association and the National Cancer Center Institute.

References

Artavanis-Tsakonas S, Matsuno K, Fortini ME (1995) Notch signaling. Science 268: 225-232
Baumeister R, Leimer U, Zweckbronner I, Jakubek C, Grünberg J, Haass C (1997) Human presenilin 1, but not familial Alzheimer's disease (FAD) mutants, facilitate *Caenorhabditis elegans* Notch signalling independently of proteolytic processing. Genes Function, in press
Borchelt DR, Thinakaran G, Eckman CB, Lee MK, Davenport F, Ratovitsky T, Prada C-M, Kim G, Seekins S, Yager D, Slunt HH, Wang R, Seeger M, Levey AI, Gandy SE, Copeland NG, Jenkins NA, Price DL, Younkin SG, Sisodia SS (1996) Familial Alzheimer's disease-linked presenilin 1 variants elevate Aβ1-42/1-40 ratio in vitro and in vivo. Neuron 17: 1005-1013
Boteva K, Vitek M, Mitsuda H, de Silva H, Xu P-T, Small G, Gilbert JR (1996) Mutation analysis of presenilin 1 gene in Alzheimer's disease. Lancet 347: 130-131
Campion D, Flaman JM, Brice A, Hannequin D, Dubois B, Martin C, Moreau V, Charbonnier F, Didierjean O, Tardieu S, Penet C, Puel M, Pasquier F, Ledoze F, Bellis G, Calenda A, Heilig R, Martinez M, Mallet J, Bellis M, Clerget-Darpoux F, Agid Y, Frebourg T (1995) Mutations of the presenilin 1 gene in families with early-onset Alzheimer's disease. Hum Mol Genet 4: 2373-2377
Chapman J, Asherov A, Wang N, Treves TA, Korczyn AD, Goldfarb LG (1995) Familial Alzheimer's disease associated with S182 codon 286 mutation. Lancet 346: 1040
Chartier-Harlin M-C, Crawford F, Houlden H, Warren A, Hughes D, Fidani L, Goate A, Rossor M, Roques P, Hardy J, Mullan M (1991) Early-onset Alzheimer's disease caused by mutations at codon 717 of the β-amyloid precursor protein gene. Nature 353: 844-846
Citron M, Westaway D, Xia W, Carlson G, Diehl T, Levesque G, Johnson-Wood K, Lee M, Seubert P, Davis A, Kholodenko D, Motter R, Sherrington R, Perry B, Yao H, Strome R, Lieberburg I, Rommens J, Kim S, Schenk D, Fraser P, St George Hyslop P, Selkoe DJ (1997) Mutant presenilins of Alzheimer's disease increase production of 42-residue amyloid β-protein in both transfected cells and transgenic mice. Nature Med 3: 67-72
Clark RF, Cruts M, Korenblat KM, He C, Talbot C, Van Broeckhoven C, Goate AM (1995) A yeast artificial chromosome contig from human chromosome 14q24 spanning the Alzheimer's disease locus AD3. Hum Mol Genet 4: 1347-1354
Conlon RA, Reaume AG, Rossant J (1995) *Notch 1* is required for the coordinate segmentation of somites. Development 121: 1533-1545

Cruts M, Backhovens H, Wang SY, Vangassen G, Theuns J, Dejonghe C, Wehnert A, Devoecht J, deWinter G, Cras P, Bruyland M, Datson N, Weissenbach J, Dendunnen JT, Martin JJ, Hendriks L, Vanbroeckhoven C (1995) Molecular genetic analysis of familial early-onset Alzheimer's disease linked to chromosome 14Q24.3. Hum Mol Genet 4: 2363–2371

Doan A, Thinakaran G, Borchelt DR, Slunt HH, Ratovitsky T, Podlisny M, Selkoe DJ, Seeger M, Gandy SE, Price DL, Sisodia SS (1996) Protein topology of presenilin 1. Neuron 17: 1023–1030

Duff K, Eckman C, Zehr C, Yu X, Prada C-M, Perez-Tur J, Hutton M, Buee L, Harigaya Y, Yager D, Morgan D, Gordon MN, Holcomb L, Refolo L, Zenk B, Hardy J, Younkin S (1996) Increased amyloid-$\beta 42(43)$ in brains of mice expressing mutant presenilin 1. Nature 383: 710–713

Goate A, Chartier-Harlin M-C, Mullan M, Brown J, Crawford F, Fidani L, Giuffra L, Haynes A, Irving N, James L, Mant R, Newton P, Rooke K, Roques P, Talbot C, Pericak-Vance M, Roses A, Williamson R, Rossor M, Owen M, Hardy J (1991) Segregation of a missense mutation in the amyloid precursor protein gene with familial Alzheimer's disease. Nature 349: 704–706

Hendricks L, van Duijn CM, Cras P, Cruts M, Van Hul W, van Harskamp F, Warren A, McInnis MG, Antonarakis SE, Martin J-J, Hofman A, Van Broeckhoven C (1992) Presenile dementia and cerebral haemorrhage linked to a mutation at codon 692 of the β-amyloid precursor protein gene. Nature Genet 1: 218–221

Hendriks L, Thinakaran G, Harris CL, De Jonghe C, Martin J-J, Sisodia SS, Van Broeckhoven C (1997) Processing of presenilin 1 in brains of Alzheimer's disease patients and controls. Neuroreport 8: 1717–1721

Hollmann M, Maron C, Heinemann S (1994) N-glycosylation site tagging suggests a three transmembrane domain topology for the glutamate receptor GluR1. Neuron 13: 1331–1343

Kim T-W, Pettingell WH, Hallmark OG, Moir RD, Wasco W, Tanzi RE (1997) Endoproteolytic cleavage and proteasomal degradation of presenilin 2 in transfected cells. J Biol Chem 272: 1106–1110

Kuner T, Wollmuth LP, Karlin A, Seeburg PH, Sakmann B (1996) Structure of the NMDA receptor channel M2 segment inferred from the accessibility of substituted cysteines. Neuron 17: 343–352

Lah JJ, Heilman CJ, Nash NR, Rees HD, Yi H, Counts SE, Levey AI (1997) Light and electron microscopic localization of presenilin 1 in primate brain. J Neurosci 17: 1971–1980

Lardelli M, Dahlstrand J, Lendahl U (1994) The novel *Notch* homologue mouse *Notch 3* lacks specific epidermal growth factor-repeats and is expressed in proliferating neuroepithelium. Mech Dev 46: 123–136

Lee MK, Slunt HH, Martin LJ, Thinakaran G, Kim G, Gandy SE, Seeger M, Koo E, Price DL, Sisodia SS (1996) Expression of presenilin 1 and 2 (PS1 and PS2) in human and murine tissues. J Neurosci 16: 7513–7525

Lemere CA, Lopera F, Kosik KS, Lendon CL, Ossa J, Saido TC, Yamaguchi H, Ruiz A, Martinez A, Madrigal L, Hincapie L, Arango JCL, Anthony DC, Koo EH, Goate AM, Selkoe DJ, Arango JCV (1996) The E280A presenilin 1 Alzheimer mutation produces increased Aβ42 deposition and severe cerebellar pathology. Nature Med 2: 1146–1150

Levey AI, Heilman CJ, Lah JJ, Nash NR, Rees HD, Wakai M, Mirra SS, Rye DB, Nochlin D, Bird TD, Mufson EJ (1997) Presenilin 1 protein expression in familial and sporadic Alzheimer's disease. Ann Neurol 47: 742–753

Levitan D, Greenwald I (1995) Facilitation of *lin-12*-mediated signalling by *sel 12*, a *Caenorhabditis elegans* S182 Alzheimer's disease gene. Nature 377: 351–354

Levitan D, Doyle TG, Brousseau D, Lee MK, Thinakaran G, Slunt HH, Sisodia SS, Greenwald I (1996) Assessment of normal and mutant human presenilin function in *Caenorhabditis elegans*. Proc Natl Acad Sci USA 93: 14940–14944

Levy-Lahad E, Wasco W, Poorkaj P, Romano DM, Oshima J, Pettingell WH, Yu C-E, Jondro PD, Schmidt SD, Wang K, Crowley AC, Fu Y-H, Guenette SY, Galas D, Nemens E, Wijsman EM, Bird TD, Schellenberg GD, Tanzi RE (1995) Candidate gene for the chromosome 1 familial Alzheimer's disease locus. Science 269: 973–977

Li X, Greenwald I (1996) Membrane topology of the C. elegans SEL-12 presenilin. Neuron 17: 1015–1021

Lindsell CE, Shawber CJ, Boulter J, Weinmaster G (1995) Jagged: a mammalian ligand that activates Notch1. Cell 80: 909–917

Mercken M, Takahashi H, Honda T, Sato K, Murayama M, Nakazato Y, Noguchi K, Imahori K, Takashima A (1996) Characterization of human presenilin 1 using N-terminal specific monoclonal antibodies: Evidence that Alzheimer mutations affect proteolytic processing. FEBS Lett 389: 297–303

Mullan M, Crawford F, Axelman K, Houlden H, Lillius L, Winblad B, Lannfelt L (1992) A pathogenic mutation for probable Alzheimer's disease in the APP gene at the N-terminus of β-amyloid. Nature Genet 1: 345–347

Naruse S, Igarashi S, Kobayashi H, Aoki K, Inuzuka T, Kaneko K, Shimizu T, Iihara K, Kojima T, Miyatake T, Tsuji S (1991) Mis-sense mutation Val-→Ile in exon 17 of amyloid precursor protein gene in Japanese familial Alzheimer's disease. Lancet 337: 978–979

Perez-Tur J, Froehlich S, Prihar G, Crook R, Baker M, Duff K, Wragg M, Busfield F, Lendon C, Clark RF, Roques P, Fuldner RA, Johnston J, Cowburn R, Forsell C, Axelman K, Lilius L, Houlden H, Karran E, Roberts GW, Rossor M, Adams MD, Hardy J, Goate A, Lannfelt L, Hutton M (1995) A mutation in Alzheimer's disease destroying a splice acceptor site in the presenilin-1 gene. Neuroreport 7: 297–301

Reaume AG, Conlon RA, Zirngibl R, Yamaguchi TP, Rossant J (1992) Expression analysis of a *Notch* homologue in the mouse embryo. Dev Biol 154: 377–387

Rogaev EI, Sherrington R, Rogaeva EA, Levesque G, Ikeda M, Liang Y, Chi H, Lin C, Holman K, Tsuda T, Mar L, Sorbi S, Nacmias B, Piacentini S, Amaducci L, Chumakov I, Cohen D, Lannfelt L, Fraser PE, Rommens JM, St George Hyslop PH (1995) Familial Alzheimer's disease in kindreds with missense mutations in a gene on chromosome 1 related to the Alzheimer's disease type 3 gene. Nature 376: 775–778

Schellenberg GD (1995) Genetic dissection of Alzheimer's disease, a heterogeneous disorder. Proc Natl Acad Sci USA 92: 8552–8559

Scheuner D, Eckman C, Jensen M, Song X, Citron M, Suzuki N, Bird TD, Hardy J, Hutton M, Kukull W, Larson E, Levy-Lahad E, Viitanen M, Peskind E, Poorkaj P, Schellenberg G, Tanzi R, Wasco W, Lannfelt L, Selkoe D, Younkin S (1996) Secreted amyloid β-protein similar to that in the senile plaques of Alzheimer's disease is increased *in vivo* by the presenilin 1 and 2 and *APP* mutations linked to familial Alzheimer's disease. Nature Med 2: 864–852

Schwab ME (1977) Ultrastructural localization of a nerve growth factor-horseradish peroxidase (NGF-HRP) coupling product after retrograde axonal transport in adrenergic neurons. Brain Res 130: 190–196

Seeger M, Nordstedt C, Petanceska S, Kovacs DM, Gouras GK, Hahne S, Fraser P, Levesque L, Czernik AJ, St George Hyslop P, Sisodia SS, Thinakaran G, Tanzi RE, Greengard P, Gandy S (1997) Evidence for phosphorylation and oligomeric assembly of presenilin 1. Proc Natl Acad Sci USA 94: 5090–5094

Sherrington R, Rogaev EI, Liang Y, Rogaeva EA, Levesque G, Ikeda M, Chi H, Lin C, Li G, Holman K, Tsuda T, Mar L, Foncin J-F, Bruni AC, Montesi MP, Sorbi S, Rainero I, Pinessi L, Nee L, Chumakov I, Pollen D, Brookes A, Sanseau P, Polinsky RJ, Wasco W, Da Silva HAR, Haines JL, Pericak-Vance MA, Tanzi RE, Roses AD, Fraser PE, Rommens JM, St George Hyslop PH (1995) Cloning of a gene bearing missense mutations in early-onset familial Alzheimer's disease. Nature 375: 754–760

Slunt HH, Thinakaran G, Lee MK, Sisodia SS (1995) Nucleotide sequence of the chromosome 14-encoded *S182* cDNA and revised secondary structure prediction. Amyloid Int J Exp Clin Invest 2: 188–190

Thinakaran G, Borchelt DR, Lee MK, Slunt HH, Spitzer L, Kim G, Ratovitski T, Davenport F, Nordstedt C, Seeger M, Hardy J, Levey AI, Gandy SE, Jenkins N, Copeland N, Price DL, Sisodia SS (1996) Endoproteolysis of presenilin 1 and accumulation of processed derivatives *in vivo*. Neuron 17: 181–190

Tomita T, Maruyama K, Saido TC, Kume H, Shinozaki K, Tokuhiro S, Capell A, Walter J, Gruenberg J, Haass C, Iwatsubo T, Obata K (1997) The presenilin 2 mutation (N141I) linked to familial Alzheimer disease (Volga German families) increases the secretion of amyloid β protein ending at the 42nd (or 43rd) residue. Proc Natl Acad Sci USA 94: 2025–2030

Wasco W, Pettingell WP, Jondro PD, Schmidt SD, Gurubhagavatula S, Rodes L, DiBlasi T, Romano DM, Guenette SY, Kovacs DM, Growdon JH, Tanzi RE (1995) Familial Alzheimer's chromosome 14 mutations. Nature Med 1: 848

Weinmaster G, Roberts VJ, Lemke G (1992) *Notch2*: a second mammalian *Notch* gene. Development 116: 931–941

Williams R, Lendahl U, Lardelli M (1995) Complementary and combinatorial pattern of Notch gene family expression during early mouse development. Mech Dev 53: 357–368

Wo ZG, Oswald RE (1994) Transmembrane topology of two kainate receptor subunits revealed by N-glycosylation. Proc Natl Acad Sci USA 91: 7154–7158

Wong PC, Zheng H, Chen H, Becher MW, Sirinathsinghji DJS, Trumbauer ME, Chen HY, Price DL, Van der Ploeg LHT, Sisodia SS (1997) Presenilin 1 is required for *Notch1* and *Dll1* expression in the paraxial mesoderm. Nature 387: 288–292

Mechanistic Studies of the Effect of Presenilins 1 and 2 on APP Metabolism

D. J. Selkoe*, W. Xia*, J. Zhang*, M. B. Podlisny*, C. A. Lemere*, M. Citron**, and E. H. Koo***

Introduction

Humans inheriting missense mutations in the presenilins (PS) 1 and 2 genes undergo progressive cerebral deposition of the amyloid β-protein (Aβ) at a very early age and develop a clinically and pathologically severe form of the familial Alzheimer's disease (FAD). Because PS1 and PS2 mutations cause the most aggressive known form of AD, it is important to elucidate the structure and function of this multi-transmembrane protein and the mechanism by which it produces disease. To these ends, we have carried out a series of studies during the last two years that provide strong evidence that mutant presenilins act by enhancing the amyloidogenic processing of the β-amyloid precursor protein (APP). Here, we will review some of the salient findings from studies of the presenilin polypeptides and their pathogenic effects from our laboratory.

PS1 Undergoes Heterogeneous Endoproteolysis Between Thr_{291} and Ala_{299} and Occurs as Stable N- and C-terminal Fragments in Normal and Alzheimer Brain Tissue

To detect and characterize the PS polypeptides in brain and other tissues, we raised antibodies to synthetic peptides or a recombinant fusion protein from several PS regions (Podlisny et al. 1997). Upon immunoprecipitation of ^{35}S-methionine-labeled transfected cells, antibodies 4627 to the C-terminus and 4624 to the putative hydrophilic loop detected full-length PS1 as two closely spaced proteins migrating at ~43–45 kDa. We always saw a characteristic band pattern: a tightly focused, sharp upper band and a broad, fuzzy lower band. Both of these bands were immunospecific, as confirmed by peptide absorptions and use of pre-immune antisera (Podlisny et al. 1997). Attempts to demonstrate a posttranslational modification that could explain the occurrence of these two apparently full-length PS1 polypeptides revealed no evidence of glycosylation or sulfation

---

* Center for Neurologic Disease, Brigham and Women's Hospital, Department of Neurology, Harvard Medical School, Boston, MA 02115-5716
** Amgen, Inc., Thousand Oaks, CA 91320-1789
*** Department of Neuroscience, University of California, San Diego, CA 92093-0691

(Xia et al. 1997). Studies by Walter et al. (1996) of the phosphorylation status of PS1 and PS2 under basal conditions demonstrated serine phosphorylation in the amino terminal portion of PS2 but no constitutive phosphorylation of PS1. In view of this lack of evidence of a posttranslational modification explaining the two PS1 bands, we carried out radiosequencing using ^{35}S-methionine. This showed that both the 43 and 45 kDa bands began at the predicted methionine$_1$ at the PS1 N-terminus (Xia et al. 1997). We concluded that the two polypeptides seen in transfected cells expressing PS1 and PS2 represent SDS-stable conformers of the PS1 full-length protein.

Under brief labeling conditions, immunoprecipitation of COS transient PS1 transfectants revealed very few or no N- and C-terminal PS1 proteolytic fragments, which have been described previously (Thinakaran et al. 1996) and which we subsequently detected in stable PS1 transfectants after longer metabolic labeling. In addition to recognizing the PS monomers, all of our PS antibodies specifically precipitated higher molecular weight (MW) bands at roughly 100, 150 and >250 kDa (Podlisny et al. 1997). The facts that the immunoprecipitated higher MW proteins can be directly immunoblotted without PS antibodies and that the ratio of their amounts to the amount of the 43–45 kDa monomers is constant at different levels of PS1 expression are consistent with the conclusion that they represent SDS-stable oligomers (see also Walter et al. 1996).

Because our initial attempts to detect endogenous PS proteins in brain and non-transfected cultured cells by immunoblotting or immunoprecipitation (i.p.) suggested that the amount of holoprotein was very low, we developed a more sensitive detection method. We covalently coupled our polyclonal antibodies 4627 (PS C-terminus) or X31 (PS1 N-terminus) to protein A Sepharose beads, immunoprecipitated PS proteins from ~1 mg of cold cell lysates (or 20–40 mg of brain or other tissues) and immunoblotted the precipitates with these and other PS1 antibodies. In PS1-transfected human 293 cells, the characteristic 43–45 kDa PS1 polypeptides (see above) were specifically detected. In the PS1-untransfected 293 cells, we specifically detected small amounts of the 43–45 kDa polypeptides, which comigrated with those overexpressed in the PS1 transfectants. The untransfected cells also contained a strongly labeled smaller protein of ~18 kDa flanked by a group of four or more weaker bands from ~17–25 kDa that were precipitated and Western blotted by PS1 antibodies. An additional smaller band at ~14 kDa (sometimes flanked by two weaker bands) was also specifically precipitated and detected by C-terminal but not by "loop" PS1 antibodies, suggesting that the smaller fragments begin after the beginning of the loop domain. Interestingly, the majority of these various low molecular weight bands were only modestly increased in amount in the PS1 transfectants compared to untransfected cells, consistent with the regulation of PS1 endoproteolysis reported by Thinakaran et al. (1996).

We also detected complementary N-terminal fragments (NTF). In non-transfected cells, our antibody X81 precipitated a strongly reactive ~28 kDa protein and a slightly larger minor band. Again, these proteins were only modestly increased in amount in PS1-transfected cells. In addition, there were several faintly labeled bands below the major 28 kDa N-terminal fragment that were not visible at all in untransfected cells.

To show directly that a precursor-product relationship exists between the holoprotein and its endoproteolytic fragments, we carried out pulse-chase experiments on PS1-transfected 293 cells. Labeling for 20 min followed by a cold chase showed that the holoprotein and the higher molecular weight aggregates gradually declined in amount with a calculated half-life of 55 minutes (Podlisny et al. 1997). The 28- and 18-kDa major proteolytic fragments were absent at time 0 and first appeared at 20 min of chase and increased modestly thereafter, up to approximately 80 min. Cells transfected with wild-type or FAD-linked mutant PS1 showed the same fragment formation, and no major difference in the pulse-chase kinetics was observed. To assess the stability of the fragments once they were formed, cells were labeled for one hour and then chased for up to six hours. In this experiment, PS holoprotein declined rapidly over one hour as expected, whereas the 28 and 18 kDa N-terminal and C-terminal fragments did not show a substantial decrease over time, even after all of the holoprotein had been catabolized. These data indicate that the endoproteolytic fragments have a half-life of well over six hours.

Radiosequencing Establishes the Sites of Endoproteolysis in PS1

To determine the exact location of the endoproteolytic cleavages that generate the fragments, we performed radiosequencing with ^{35}S-methionine or ^{3}H-valine on the major 18 kDa C-terminal fragment immunoprecipitated from PS1-transfected human 293 cells. The positions of radioactive methionine and valine in the Edman degradation clearly indicated that the major site of endoproteolysis in these cells was between residues 298 and 299 of PS1 and minor sites were observed between 291 and 292 and between 292 and 293. Therefore, even the major C-terminal immunoreactive fragment detected as a single radioactive gel band in these cells shows N-terminal heterogeneity. Because there are other N- and C-terminal immunoreactive fragments flanking the major bands, it is clear that PS1 (and presumably PS2) undergoes considerable heterogeneity of endoproteolysis.

Detection of Presenilin Proteins in Monkey and Human Brain Tissue

Based on the above characterization of PS1 and its fragments in 293 and other cells, we used combined i.p./Western blotting to substantially enrich PS proteins from fresh frozen Rhesus monkey and mouse brains and from AD, FAD and control postmortem human brains. In the case of the fresh animal brain, amounts of starting tissue as high as 40 mg wet weight failed to reveal any specifically reactive holoprotein in the 40–50 kDa range. Instead, we consistently detected a major 18 kDa C-terminal fragment (CTF) and minor flanking proteins. Blotting with C-terminal PS1 antibodies revealed the additional 14 kD fragment (see above) that apparently lacks the "loop" region. These studies also revealed the major 28 kDa N-terminal fragment and two to three faint, slightly larger bands.

Similar experiments on homogenates of frozen human cerebral cortex again failed to reveal definite, full-length PS polypeptides, despite the large amounts (20–40 mg) of starting tissue used for i.p./Western blotting. The NTF and CTF patterns were similar to those of monkey brain, except that in most human brains, a ~35 kDa NTF and a 14 kDa CTF were detected at similar intensities to the 28 kDa N-terminal and 18 kDa C-terminal bands. Thus, adult human brain appears to have substantial amounts of these alternative endoproteolytic fragments in cortex, and these must arise from a cleavage somewhat C-terminal to the usual 298/299 position. To examine whether FAD-linked missense mutations in PS1 alter the fragment pattern *in vivo*, we obtained postmortem brain (courtesy of Dr. D. Pollen, University of Massachusetts, Worcester) from four different patients of the FAD family harboring the cys410tyr PS1 mutation. Although some sample to sample variation was observed in these postmortem brains, we saw no convincing differences in the multiple C- and N-terminal fragments of PS among the four PS1 FAD brains, four sporadic AD brains and four control brains. No full-length PS polypeptides were specifically detected in any of these 12 human brains.

Mutations in PS1 and PS2 Selectively Increase Aβ_{42} Production in Transfected Cells

We have carried out an extensive series of studies on both 293 and CHO cells transfected with wild-type or several different mutant PS1 or PS2 cDNAs (Xia et al. 1997; Citron et al. 1997). Regardless of the relative level of PS1 or PS2 expression in these various transfectants, all of which also expressed transfected full-length APP (either the wild-type or Swedish isoforms), we did not observe any consistent change in the levels of APP holoprotein in the lysate or APPs in the medium as a result of expressing the mutant forms. However, both gel phosphorimaging of immunoprecipitated Aβ_{42} species (Xia et al. 1997) and by sensitive sandwich ELISA detection of Aβ_{42} and Aβ_{total} peptides (Xia et al. 1997; Citron et al. 1997) consistently revealed a 1.5–4.0-fold elevation in Aβ_{42}/Aβ_{total} ratios in the conditioned media of the cell lines expressing mutant PS1 or mutant PS2 compared to wild-type transfectants. The relative elevation of Aβ_{42} occurred independently of the level of expression of the mutant PS1 or PS2 proteins; even minimal overexpression of mutant presenilin led directly to elevation of Aβ_{42} to levels similar to those observed in much higher expressing mutant transfectants. Although occasional, modest increases in Aβ_{total} levels were observed in some clonal lines, this was not consistent among all lines expressing a particular mutation non among all the PS mutations studied. We conclude that there is a reproducible and highly statistically significant selective elevation of Aβ_{42} production in different cell lines expressing different PS1 or PS2 mutant molecules. The fact that this effect occurs in non-neural (293) and even non-human (CHO) cells in the absence of any neural or other AD-type influence supports the hypothesis that $A\beta_{42}$ elevation is a direct and primary effect of presenilin mutations and does not arise secondary to other pathogenic effects that such mutations might hypothetically induce in the course of AD.

Similar results in PS transfected cells have been seen in other laboratories (Borchelt et al. 1996) and have been confirmed in transgenic mice expressing mutant presenilins in several laboratories (Citron et al. 1997; Borchelt et al. 1996; Duff et al. 1996).

To obtain *in vivo* information confirming this effect of PS1 mutations on the 42-specific γ-secretase processing of APP, we characterized the neuropathological phenotype of four PSA1 FAD patients from a large Columbian kindred bearing the codon 280 glu to ala substitution Lemere et al. 1996. Using antibodies specific to the alternative C-termini of Aβ (40 and 42), we detected massive deposition of $A\beta_{42}$, the earliest and predominant plaque Aβ to isoform occur in AD, in many brain regions. Computer-assisted quantification revealed a significant increase in $A\beta_{42}$, but not $A\beta_{40}$, burden in the brains of the four PS1-FAD patients compared with those of 12 sporadic AD patients of similar neuropathological severity. Severe cerebellar pathology included numerous $A\beta_{42}$-reactive plaques, many bearing dystrophic neurites and reactive glia. Our results in brain tissue are entirely consistent with the biochemical evidence we and others have obtained of increased $A\beta_{42}$ levels in PS1-transfected cells, transgenic mice and FAD plasma and fibroblast media samples. Taken together, the results strongly suggest that mutant PS1 proteins selectively alter the proteolytic processing of APP at the C-terminus of Aβ to favor production of $A\beta_{42}$.

Evidence of In Vivo Interaction Between Amyloid Precursor Protein and Presenilins in Mammalian Cells

The mechanism by which mutations in PS1 and PS2 lead to the selective increase in $A\beta_{42}$ production documented above is unknown. Because both presenilin and APP are expressed in the endoplasmic reticulum (ER) and Golgi apparatus (Walter et al. 1996; Kovacs et al. 1996; Cook et al. 1996; De Strooper et al. 1997), it is possible that a direct or indirect interaction of the two proteins in these compartments could underlie the effect of mutant presenilin on APP proteolysis. To address this hypothesis, we have systematically searched for evidence of APP-PS complexes in both non-human (CHO) and human (293 and HS683) cells.

We first used a variety of CHO cell lines stably expressing human APP and/or PS1 or PS2 protein (Xia et al. 1997). Some cells were solely transfected with APP and expressed only endogenous levels of PS, whereas others were doubly transfected. We also examined some cell lines that were transfected with PS1 alone and contained only endogenous levels of APP. To search for PS-APP complexes, we again used a highly sensitive i.p.-Western blot technique (Podlisny et al. 1997). In cells expressing transfected human APP and endogenous PS1 (our cell line 7W), we found that APP co-precipitated with PS1, as detected by the APP monoclonal antibody 8E5. APP could also be co-precipitated by our high-titer PS1 antibody, X81, in a cell line ($PS1_{Wt-2}$) that expresses transfected APP and a low level of transfected PS1 (but still higher than endogenous PS1). A large amount of APP was co-precipitated by X81 antibody in our $PS1_{Wt-1}$ line, which expresses high levels of both PS1 and APP. The amounts of APP-PS1 complexes detected in these various cell lines were always directly related to the levels of expression of the two proteins, suggesting a specific interaction. Highly similar results were obtained with several different PS1 and APP antibodies. The APP molecules co-precipitated with PS1 were almost solely the N-glycosylated form; very little N+O-glycosylated APP came down with PS1 in the cell lysates.

Next, we showed that the same type of complex formation could occur between APP and PS2. Three different PS2 antibodies were used for i.p., followed by Western blotting with 8E5, again showing consistent co-i.p. of APP and PS2.

Next, we reversed the sequence of antibodies, namely precipitating with APP and then blotting with PS2 antibodies. This procedure resulted in detection of the characteristic thin and broad PS1 holoprotein bands at 43–45 kDa. We also searched for APP-PS1 complex formation in another cell type, human 293 cells; again, N-glycosylated APP was precipitated with PS1. In this case, the 293 cells were transfected with the 695-residue isoform, and this migrated at the expected position after co-i.p., just as the 751 form of APP had done in the CHO transfectants.

To confirm unequivocally the physiological formation of APP-PS complexes at endogenous protein levels, we used X81 or 4627 (to the PS1 N- or C-termini) to precipitate a large amount of lysate from non-transfected CHO cells and then blotted with 22C11, an antibody recognizing both human and hamster APP. We were able to detect N-glycosylated APP after PS1 i.p. The same result was obtained in a CHO line transfected solely with PS2 without APP. These results clearly demonstrated that APP-PS complexes existed at endogenous levels without the need for overexpression by transfection. Further, we detected endogenous APP-PS complexes in non-transfected human 293 cells and human HS683 glioma cells.

We analyzed complex formation in a stable CHO cell line expressing APP_{751} truncated at residue 709 and thus lacking almost the entire cytoplasmic domain (710–751). We observed a co-precipitating APP species that migrated just below the full-length N-glycosylated APP and co-migrated precisely with the truncated N-glycosylated APP species precipitated by an APP antibody (B5) from the same cell lysate. Therefore, it appears that the cytoplasmic domain of APP is dispensable for the interaction with PS1.

We next demonstrated APP-PS co-immunoprecipitation in cells expressing the APP KN651/652NL (Swedish) or V698F (Indiana) missense mutations in APP_{751}. There was no obvious change in the ability to co-i.p. APP using PS antibodies. Similarly, we performed such experiments on two cell lines expressing PS1 mutations (M146L and C410Y) and two lines expressing PS2 mutations (the Volga German and Italian mutations). In all these cases, the characteristic N-glycosylated APP band was detected after precipitation of lysates with PS antibodies. No consistent qualitative or quantitative differences in complex formation between wild-type and mutant proteins have been detected to date using the current methods.

Because PS1 and PS2 proteins have been localized principally to the ER, we expected the formation of APP-PS complexes to occur at least in part in this compartment. To assess this, double transfectants were treated with brefeldin A (BFA) for one hour followed by i.p./Western blotting, as above. BFA, which causes a collapse of the Golgi and retention of proteins in the ER, results in a change in APP glycosylation, with one broad band of intermediate size. Upon i.p. with PS1 or PS2 antibodies, this intermediate APP band was clearly detected by 8E5 Western blotting. Similarly, a 20°C temperature block, which results in protein retention principally in the trans Golgi network, did not obviate APP-PS co-precipitation. These results suggest that the ER is a principal compartment in which the APP-PS interaction occurs. We have recently carried out subcellular fractionation on CHO APP/PS double transfectants. Calnexin has been used as a marker for ER vesicles, while galactosyl transferase has been used as an enzymatic assay for Golgi vesicles. In discontinuous sucrose density gradients, we obtained ER and Golgi fractions containing these respective markers. In addition, the putative Golgi fractions contained the N+O-glycosylated APP, whereas the ER fractions contained only N-glycosylated APP, as expected. We performed PS i.p. followed by APP Western blotting and were able to detect the characteristic N-glycosylated APP band in the PS immunoprecipitates of both ER and Golgi fractions. In the case of the Golgi fractions, we obtained a small amount of N+O-glycosylated APP as well. We then performed these studies with both wild-type and mutant PS1 or PS2 transfectants and have to date seen no clear difference; all cell lines show an ability to co-precipitate PS and APP in ER and Golgi fractions. It is the PS1 and PS2 holoproteins that are precipitated with APP. Further studies will be needed to determine to what extent the PS N- and C-terminal fragments may participate in the complex; we have no evidence about this at this writing.

Discussion

Our studies demonstrate several features of presenilin protein expression in cell lines and in animal and human brains. Among the various findings, we have established residues 298–299 as the principal site for endoproteolytic cleavage generating the major C-terminal fragment of PS1 in human cells.

Extensive studies using a variety of cell lines demonstrate a selective increase in $A\beta_{42}/A\beta_{total}$ ratios in cells expressing mutant PS1 or PS2. These results are reflected by a statistically significant two to three fold increase in $A\beta_{42}$ plaque burden in four PS1 (E280A) brains compared to sporadic AD and control brains.

We have begun to address the mechanism by which mutant presenilins alter APP γ-secretase processing by searching for and detecting an apparent interaction between APP and PS proteins in intact, living cells of several types. Our experiments demonstrate a highly reproducible interaction between each of the presenilins and the N-glycosylated form of APP (and to some extent, the N+O-glycosylated form), the specificity of which is supported by several lines of evidence. 1) Numerous antibodies recognizing different epitopes in PS1, PS2 or APP could all immunoprecipitate the PS-APP complexes. 2) PS and APP antibodies could be used for immunoprecipitation-Western blotting in either order with the same positive results. 3) The amount of PS-APP complexes varied directly with expression levels in stable transfectants with widely varying PS and APP levels. 4) Co-expression of a control protein, human transferrin receptor, did not lead to association with PS1, mitigating against the possibility that APP-PS interaction is simply due to overexpression in our transfectants. 5) Non-transfected cells showed PS-APP complex formation at endogenous protein levels. 6) Treatment of cells with BFA changed the glycosylation of APP, as previously reported (Haass et al. 1995), and this modified form was now specifically immunoprecipitated with PS1. 7) Stable expression of different sized APP polypeptides (APP_{695}, APP_{751} and $APP_{\Delta C}$) always led to specific co-immunoprecipitation of the appropriately sized species with PS. 8) PS1 and PS2 each underwent highly similar complex formation with APP. 9) The findings were confirmed in two distinct cell types. We conclude that PS and APP interact directly in intact living cells to form stable complexes.

It should be emphasized that none of our results exclude the possibility that additional protein constituents are present in complexes that contain APP and PS molecules in cells. Importantly, the complexes have been demonstrated directly in ER and Golgi vesicles after subcellular fractionation, in agreement with the localization of each of these molecules to these compartments in previous immunocytochemical studies. Early results suggest the possibility that the transmembrane regions of each molecule are involved in mediating the interaction, although additional studies will be needed to prove this and to learn whether the ectodomain of the proteins also play a role in the interaction. In the case of PS1, there is a relatively small ~40-residue lumenal loop between TM1 and TM2, according to current models of topology.

We have recently begun to perform intracellular ELISA measurements using sensitive sandwich assays on the same subcellular fractions in which APP-PS complexes have been detected. These results suggest that mutant PS1 and PS2 elevate $A\beta_{42}/A\beta_{total}$ ratios in Golgi vesicles of CHO double transfected cells. There is also evidence of some increase in ER-derived vesicles. Further studies are underway to confirm and extend these preliminary data. Our results suggest that mutant presenilins can alter APP proteolysis as early as the ER or early Golgi compartments. The heightened amount of intracellular $A\beta_{42}$ production apparently leads to excess secretion of this species, as its levels are clearly increased in plasma and primary fibroblast media of patients harboring PS1 or PS2 mutations (Scheuner et al. 1996).

In summary, we believe our findings provide a central clue for understanding the mechanism by which mutant PS proteins selectively alter APP processing. It appears that each of the presenilins and APP can form complexes *in vivo*, predominantly in the ER but also in the Golgi, and that the transmembrane regions may be implicated in this process. These data establish a model that may explain the $A\beta_{42}$ phenotype associated with the expression of mutant PS1 and PS2, with attendant implications for the formation of diffuse $A\beta_{42}$ plaques as an early event in Alzheimer's disease in general.

Acknowledgements. We thank members of our laboratory for contributions to these studies and for many helpful discussions. The reagents for the $A\beta$ ELISA were generously provided by P. Seubert, R. Motter and D. Schenk at Athena Neurosciences, South San Francisco, CA. This work was supported by NIH grants AG06173, AG05134 and AG12749 (to D. J. S.), AG12376 and NS 01812 (to E. H. K.), and the Foundation for Neurologic Diseases.

References

Borchelt DR, Thinakaran G, Eckman CB, Lee MK, Davenport F, Ratovitsky T, Prada C-M, Kim G, Seekins S, Yager D, Slunt HH, Wang R, Seeger M, Levey AI, Gandy SE, Copeland NG, Jenkins NA, Price DL, Younkin SG, Sisodia SS (1996) Familial Alzheimer's disease-linked presenilin 1 variants elevate $A\beta$1-42/1-40 ratio in vitro and in vivo. Neuron 17: 1005–1013

Citron M, Westaway D, Xia W, Carlson D, Diehl TS, Levesque G, Johnson-Wood K, Lee M, Seubert P, Davis A, Kholodenko D, Motter R, Sherrington R, Perry B, Yao H, Strome R, Lieberburg I, Rommens J, Kim S, Schenk D, Fraser P, St George-Hyslop P, Selkoe DJ (1997) Mutant presenilins of Alzheimer's disease increase production of 42-residue amyloid β-protein in both transfected cells and transgenic mice. Nature Med 3: 67–72

Cook DG, Sung JC, Golde TE, Felsenstein KM, Wojczk BS, Tanzi RE, Trojanowski JQ, Lee VM-Y, Doms RW (1996) Expression and analysis of presenilin 1 in a human neuronal system: localization in cell bodies and dendrites. Proc Natl Acad Sci USA 93: 9223–9228

De Strooper B, Beullens M, Contreras B, Levesque L, Craessaerts K, Cordell B, Moechars D, Bollen M, Fraser P, St. George-Hyslop P, Van Leuven F (1997) Phosphorylation subcellular localization, and membrane orientation of the Alzheimer's-disease associated presenilins. J Biol Chem 272: 3590–3598

Duff K, Eckman C, Zehr C, Yu X, Prada C-M, Perez-Tur J, Hutton M, Buee L, Harigaya Y, Yager D, Morgan D, Gordon MN, Holcomb L, Refolo L, Zenk B, Hardy J, Younkin S (1996) Increased amyloid-β42(43) in brains of mice expressing mutant presinilin 1. Nature 383: 710–713

Haass C, Lemere CA, Capell A, Citron M, Seubert P, Schenk D, Lannfelt L, Selkoe DJ (1995) The Swedish mutation causes early-onset Alzheimer's disease by β-secretase cleavage within the secretory pathway. Nature Med 1: 1291–1296

Kovacs DM, Fausett HJ, Page KJ, Kim T-W, Mori RD, Merriam DE, Hoillister RD, Hallmark OG, Mancini R, Felsenstein KM, Hyman BT, Tanzi RE, Wasco W (1996) Alzheimer-associated presenilins 1 and 2: neuronal expression in brain and localization to intracellular membranes in mammalian cells. Nat Med 2: 224–229

Podlisny MB, Citron M, Amarante P, Sherrington R, Xia W, Zhang J, Diehl T, Levesque G, Fraser P, Haass C, Koo EHM, Seubert P, St. George-Hyslop P, Teplow DB, Selkoe DJ. (1997) Presenilin proteins undergo heterogeneous endoproteolysis between Thr_{291} and Ala_{229} and occur as stable N- and C-terminal fragments in normal and Alzheimer brain tissue. Neurobiol Disease 3: 325–337

Scheuner D, Eckman C, Jensen M, Song X, Citron M, Suzuki N, Bird TD, Hardy J, Hutton M, Kukull W, Larson E, Levy-Lahad E, Viitanen M, Peskind E, Poorkaj P, Schellenberg G, Tanzi R, Wasco W, Lannfelt L, Selkoe D, Younkin S (1996) Secreted amyloid β-protein similar to that in the senile plaques of Alzheimer's disease is increased *in vivo* by the presenilin 1 and 2 and *APP* mutations linked to familial Alzheimer's disease. Nature Med 2: 864–870

Thinakaran G, Borchelt DR, Lee MK, Slunt HH, Spitzer L, Kim G, Ratovitsky T, Davenport F, Nordstedt C, Seeger M, Hardy J, Levey AI, Gandy SE, Jenkins NA, Copeland NG, Price DL, Sisodia SS

(1996) Endoproteolysis of presenilin 1 and accumulation of processed derivatives *in vivo*. Neuron 17: 181–190

Walter J, Hung AY, Sisodia SS, Selkoe DJ, Haass C (1996) Phosphorylation of β-amyloid precursor protein at two distinct cellular locations. J Biol Chem 272: 1896–1903

Xia W, Zhang J, Kholodenko D, Citron M, Podlisny MB, Teplow DB, Haass C, Seubert P, Koo EH, Selkoe DJ (1997) Enhanced production and oligomerization of the 42-residue amyloid β-protein by Chinese hamster ovary cells stably expressing mutant presenilins. J Biol Chem 272: 7977–7982

Presenilin 2 – APP Interactions

W. Wasco*, R. E. Tanzi, R. D. Moir, A. C. Crowley, D. E. Merriam,
D. M. Romano, P. D. Jondro, and B. A. Kellerman

Introduction

Familial Alzheimer's Disease Genes

Alzheimer's disease (AD) is a progressive neurodegenerative disorder of the central nervous system which is invariably associated with and defined by the presence of intracellular neurofibrillary tangles (NFT) and extracellular deposits of amyloid (Aβ senile plaques) in the brain and cerebral blood vessels. While the etiologic events that lead to the generation of these pathological hallmarks and ultimately to synaptic loss and neurodegeneration are not well understood, it is clear that a significant portion of AD has a genetic basis (see Wasco and Tanzi 1996 for review). These familial forms of Alzheimer's disease (FAD) can be classified based on both the age of onset and the type of gene defect inherited. A large proportion of early onset (<60 years old) FAD is attributed to "causative" defects in one of three genes. Mutations in the amyloid β protein precursor (APP) gene located on chromosome 21 (Goate et al. 1991; Levy et al. 1990; Murrell et al. 1991; Chartier-Harlin et al. 1991; Hendriks et al. 1992; Mullan et al. 1992) cause a small percentage of early-onset FAD, but the majority of early-onset FAD is known to be caused by mutations in two recently identified genes, presenilin 1 (PS1) and presenilin 2 (PS2); Sherrington et al. 1995; Levy-Lahad et al. 1995; Rogaev et al. 1995). An allele of a fourth gene, APOE4 which is located on chromosome 19, confers increased „risk" for late onset FAD (>60 years of age; Pericak-Vance et al. 1991; Strittmatter et al. 1993; Corder et al. 1993; Rebeck et al. 1993).

Presenilin Gene Identification and Characterization

PS1 and PS2 were both isolated in the latter half of 1995 (Sherrington et al. 1995; Levy-Lahad et al. 1995; Rogaev et al. 1995). Mutations in these two genes together with those in APP appear to account for the majority of early-onset FAD. Since its discovery in 1987, much has been learned about the processing and regulation of

---

* Genetics and Aging Unit Department of Neurology Massachusetts General Hospital-East Boston, Massachusetts, 02129-9142, USA

APP, which is a single transmembrane domain cell surface protein that matures though the intracellular compartments of the endoplasmic reticulum (ER) and Golgi (see Buxbaum and Greengard 1996 and Gandy and Greengard 1994 for review). In contrast, relatively little is known about the presenilin proteins. The presence of six to nine hydrophobic domains indicates that both presenilins share a serpentine membrane spanning topology (Sherrington et al. 1995; Levy-Lahad et al. 1995; Rogaev et al. 1995). Although the actual number of transmembrane domains (TMD) remains unclear, an eight TMD domain model appears best substantiated at this time (Lehmann et al. 1997; Doan et al. 1996; Li and Greenwald 1996). As depicted in Figure 1, the eight TMD model dictates that both proteins have one large and five smaller hydrophilic loops and that the N-terminus, the large loop and the C-terminus are located on the cytoplasmic side of the membrane. The fact that two specific regions of PS1 and PS2, the first 80 amino acids and the single large hydrophilic loop, are not particularly well conserved (either among the two mammalian proteins or in two *C. elegans* homologues; Levitan and Greenwald 1995; L'Hernault and Arduengo 1992) raises the possibility that these regions impart specificity of function or localization to the different family members, and that to do so these domains may interact with proteins located in the cytoplasm.

The normal biological role(s) of the presenilins and the mechanism(s) by which the FAD-associated mutations exert their effect remain unknown. We have shown that both proteins localize primarily to the intracellular membranes of the ER/Golgi (Kovacs et al. 1996) and that, although both proteins appear to be ubiquitously expressed, PS2 is less abundant than PS1. These findings have now been confirmed by a number of other groups. To date, over 30 separate causative mutations have been identified in PS1 and two such mutations have been detected in PS2 (see Wasco and Tanzi 1996; Hardy 1997 for reviews). All but one

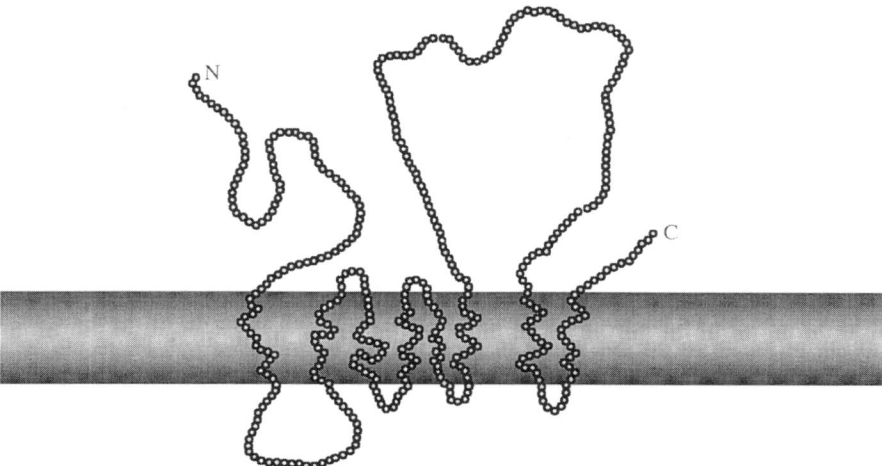

Fig. 1. Eight transmembrane domain model of the presenilins

of the identified PS alterations produce missence mutations that result in single amino acid changes. The solitary exception is a PS1 mutation that leads to an inframe deletion of exon 9 (Perez-Tur et al. 1996). Interestingly, PS1 and PS2 both contain sites for proteolytic cleavage located within the large loop predicted to be located between TMD-6 and TMD-7, and for PS1 this cleavage site is predicted to be within or near exon 9. Processing at this site results in the production of an N-terminal fragment of approximately 28 (PS1) or 30 (PS2) kDa and a C-terminal fragment of approximately 19 (PS1) or 25 (PS2) kDa (Thinakaran et al. 1996; Kim et al. 1997). We have recently demonstrated that, under certain conditions, PS2 is cleaved at an alternative site within the large loop to produce and N-terminal fragment of 35 kDa and C-terminal fragment of 20 kDa (Kim et al. 1997). Additionally, we found that the N141I PS2 mutation appears to lead to inefficient degradation of the alternative C-terminal fragment and that the accumulation of this fragment in the detergent-resistant fractions could be equalized in mutant and wild-type PS2 cell lines by blocking the proteasome (Kim et al. 1997). Taken together, these findings indicate that PS2 carrying the N141I FAD mutation is not efficiently degraded by the proteasome pathway which in turn leads to an enhanced accumulation of the alternative C-terminal cleavage product in the detergent-resistant cellular fraction. These data are described in detail elsewhere in this volume by Tanzi et al. At the present time, little is known about the regulation or function of the regulated or alternative PS1/PS2 cleavage events or about the effect that this processing has on the normal function of the proteins.

Clues to the Normal Biological Roles of the Presenilins

Clues about the function of the presenilins in mammalian cells can be derived from the observation that both PS1 and PS2 share significant amino acid homology with two *C. elegans* proteins, *sel-12* (Levitan and Greenwald 1995) and *spe-4* (L'Hernault and Arduengo 1992). It has been proposed that *sel-12* plays a role in receptor trafficking, localization or recycling of *lin-12* (Levitan and Greenwald 1995) and that *spe-4* is involved in cytoplasmic partitioning of proteins (L'Hernault and Arduengo 1992); thus the display of homology between PS1 and PS2 with the *C. elegans* proteins suggest that the presenilins may also be involved in intracellular trafficking or localization of proteins. Interestingly, as described above, it has been demonstrated that mutations in APP lead to increases in the amounts of Aβ and it has long been hypothesized that this is a result of altered intracellular trafficking and/or sorting of APP. This hypothesis gains further support from the observation that the levels of Aβ42 are increased in fibroblasts from patients with specific mutations in PS1 or PS2 (Scheuner et al. 1996).

Another clue to the normal role of the presenilins may be provided by the observation that a truncated PS2 protein produced by a clone containing the nucleotides that encode the C-terminal 103 amino acids of PS2 appears to inhibit T-cell receptor- and Fas-induced apoptosis (Vito et al. 1996). In addition, PS2 was found to confer increased sensitivity to apoptosis in PC12 cells after challenge with staurosporine or hydrogen peroxide (Deng et al. 1996). In a more recent collaborative effort we have found that transfection of NGF-differentiated PC12 cells with PS2 increased apoptosis induced by withdrawal from trophic factors, or by the addition of Aβ to the culture medium, while transfection with antisense PS2 constructs rescued cells from apoptosis (Wolozin et al. 1996). These findings suggest that overexpression of PS2 increases the susceptibility of neurons to apoptotic stimuli. This putative link between programmed cell death and one of the genes known to be involved in the etiology of Alzheimer's disease is intriguing, given that it has been suggested that programmed cell death may play a role in neurodegenerative disease, perhaps by imparting slight but pathological alterations in the controlled removal of damaged neurons.

Interaction of PS2 and APP

It is clear that individuals carrying the PS2 N141I mutation have increased β-amyloid deposition in the brain; in addition, it has recently been demonstrated that fibroblasts from patients carrying the Volga German N141I PS2 mutation secrete levels of Aβ 1-42 that are greater than those secreted by fibroblasts from unaffected controls (Scheuner et al. 1996; Citron et al. 1997; Borchelt et al. 1996). To date, it is not known if the N141I-associated increase in Aβ deposition is due to a direct interaction between PS2 and APP or to downstream effects of PS2 mutations on APP trafficking and processing. APP can be processed via at least two different types of pathways: those that preclude Aβ formation and those that encourage it, and given the pathological consequences of overproduction of Aβ, it is probable that the equilibrium between these pathways is tightly maintained. Possible points of perturbation in the pathway include factors and processes involved in the localization and/or trafficking of APP. Current hypotheses indicate that the FAD mutations can lead to the presence of APP in cellular compartments that are more conductive to the production of Aβ peptide or longer forms of Aβ (Aβ1-42) that are more amyloidogenic. In fact, data currently exist to show that mutations in APP can result in a shunt of the molecule to more amyloidogenic pathway(s). In cells of unaffected individuals, this amyloidogenic pathway may only be used occasionally, e.g., if the non-amyloidogenic pathway(s) is overburdened or disabled. The location of all of the known APP FAD mutations within or around the cleavage site for a secretase indicates that they most likely cause FAD by affecting the processing of APP and generation of Aβ (Cai et al. 1993; Citron et al. 1993; Suzuki et al. 1994). In addition, results from a number of studies indicate that the FAD-associated mutations in APP directly alter their normal trafficking/localization profiles (for review, see Mellman et al. 1995; Selkoe 1995). Although it is clear that the N141I Volga German mutations in PS2 are linked to an increase in the levels of Aβ, whether this observed increase is due to direct interaction between APP and PS2 or to an interaction with other proteins in the Aβ-producing pathway (for instance α and/or β secretase enzymes) remains unclear.

Co-Immunoprecipitation of APP and PS2

As a first step in assessing the ability of APP and PS2 to directly interact, we carried out experiments to determine whether antibodies that immunoprecipitate PS2 will also immunoprecipitate APP. Given that the expression of endogenous PS2 is low, we have initially made use of H4 cells that have been transfected with wild type or N141I mutant PS2 containing an inducible promoter regulated by a tetracycline-repressible transactivator (Gossen and Bujard 1992; Kim et al. 1997). Protein extracts from uninduced control cells or from cells that were induced to produce epitope-tagged wild type or N141I PS2, were prepared and immunoprecipitated with an antibody that recognizes the epitope tag (M2), or with a control antisera. Western blots using antibodies that detect APP demonstrate that APP is present in the samples immunoprecipitated with antisera that recognize PS2, but not in the samples containing the control antisera (Fig. 2). The co-immunoprecipitation of APP was entirely dependent on the presence of PS2, as evidenced by the inability of APP to be immunoprecipitated from extracts derived from cells that were not induced to produce PS2 (data not shown). This finding indicates that APP and PS2 are able to be co-immunoprecipitated and suggests that the two proteins may interact *in vivo*. To further test this interaction we immunoprecipitated induced or uninduced cell extracts with an antibody that recognizes APP, and carried out Western blot analysis with an antibody that recognizes the epitope-tagged PS2 (Fig. 2). We found that PS2 was able to be co-immunoprecipitated with APP, but not with the control antisera. In both immunoprecipitation paradigms, which are carried out in the presence of detergents and thus involve the solubilization of proteins from intracellular membranes, we did not detect a qualitative or quantitative difference in the ability of mutant vs. wild type PS2 to be co-immunoprecipitated with APP.

Chemical Crosslinking Analysis of Wild Type and Mutant PS2 Interaction with APP in Intact, Viable Cells

To test for an *in vivo* association between PS2 and APP, we subjected the PS2-inducible cells, while still growing on tissue culture plates, to a cell membrane-permeable chemical crosslinker, disuccimydylsuberate (DSS). After incubation with DSS, extracts from wild-type or mutant PS2 cells were prepared and then immunoprecipitated with an antibody that recognizes the epitope-tagged PS2 (M2), separated by SDS-PAGE and analyzed by Western blot with antibodies that recognize APP (22C11 and C7). The results of this analysis are shown in Figure 3. After incubation in the presence of crosslinker, immunoprecipitation with the M2 antibody results in the immunoprecipitation of a complex that runs at >200 kDa and is recognized by two different anti-APP antibodies..This finding indicates that a complex containing both APP and PS2 can be immunoprecipitated by M2. This complex was not detected in extracts derived from cells that were incubated in the absence of crosslinker. Similar results were obtained when an anti-APP

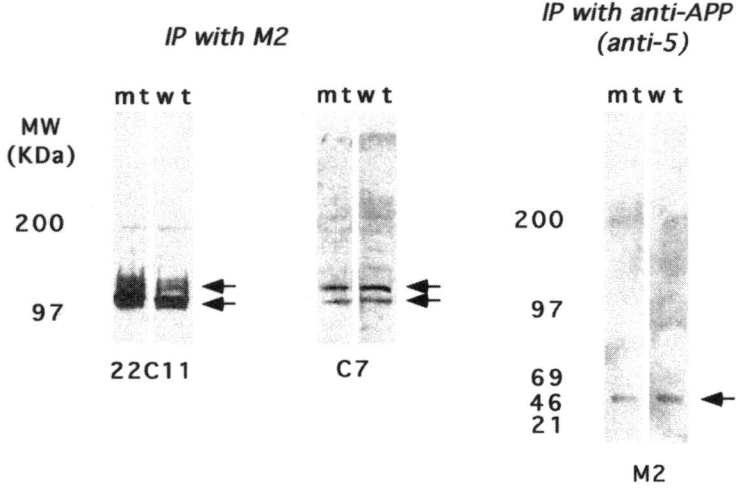

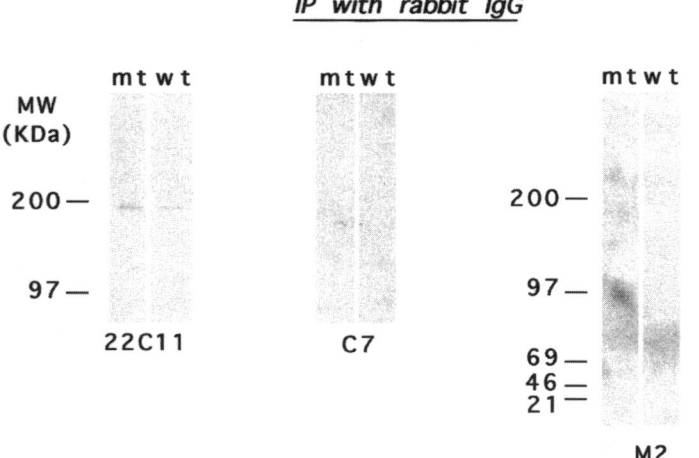

Fig. 2. Coimmunoprecipitation of AP and PS2. Extracts from cells induced to produce wild-type (wt) N141I PS2 (mt) PS2 were immunoprecipitated with the antibodies as indicated above the panels, and then analyzed by Western blot with the antibody indicated below the panel. Arrows indicate the position of the mature and immature forms of APP detected by 22C11 and C7 at approximately 110-120 kDa and of PS2 detected by M2 at approximately 54 kD (upper panel). Neither APP nor PS2 was immunoprecipitated by a control antiserum (rabbit IgG; lower panel)

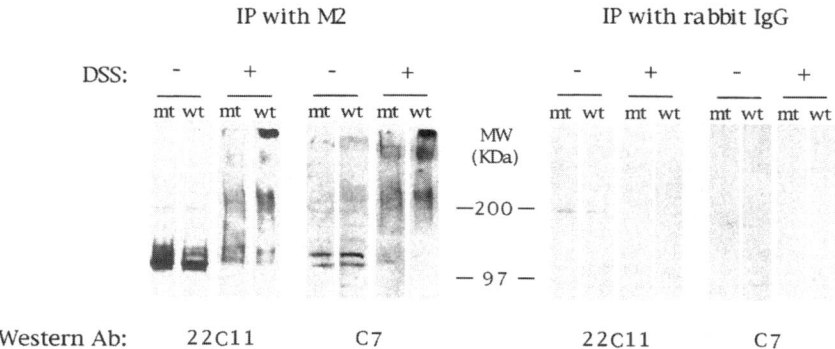

Fig. 3. In vivo crosslinking of APP and PS2. Human h4 neuroglioma cells induced to produce wild type (wt) or mutant (mt) PS2 were incubated in the presence (+) or absence (-) of crosslinker (DSS). After neutralization of the crosslinking reagent, cellular extracts were prepared, immunoprecipitated with either M2, which recognizes the induced epitope tagged forms of PS2, or with rabbit IgG. Immunoprecipitated protein was separated by SDS-PAGE and analyzed by Western blot with the indicated antibody

antibody (anti-5) was used for immunoprecipitation and M2 was used for detection (data not shown). Notably, the higher molecular weight complexes were consistently considerably more abundant in the cells expressing wild-type as opposed to mutant PS2. No immunoreactive material was apparent in the control samples that were immunoprecipitated with rabbit IgG.

Modified ELISA Assay

To directly measure the ability of PS2 to bind to APP, we used a modified enzyme-linked immunosorbant assay (ELISA) in which secreted APP (APPs) is captured on zinc(II) coated microtiter wells (Bush et al. 1994). To measure the ability of PS2 to interact with the APPs bound to these plates, was prepared partially purified epitope-tagged PS2 from human H4 neuroglioma cells induced to produce PS2. To measure binding of PS2 to the captured APPs, increasing amounts of PS2 were added to the microtiter wells and the amount of PS2 bound to the microtiter dish was measured using a monoclonal antibody specific for the PS2 produced under inducing conditions. Binding of the antibody to APPs containing wells, but not to wells in which APPs was omitted, confirms that APPs was effectively captured by the metal-coated microtiter wells. Figure 4 shows the dose response for PS2 binding to APPs captured on the microtiter wells. PS2 did not bind to wells in which APPs was omitted; neither was any significant binding detected when extracts from uninduced H4 cells was added to the APP containing wells. These results provide direct evidence for an interaction between APPs and PS2.

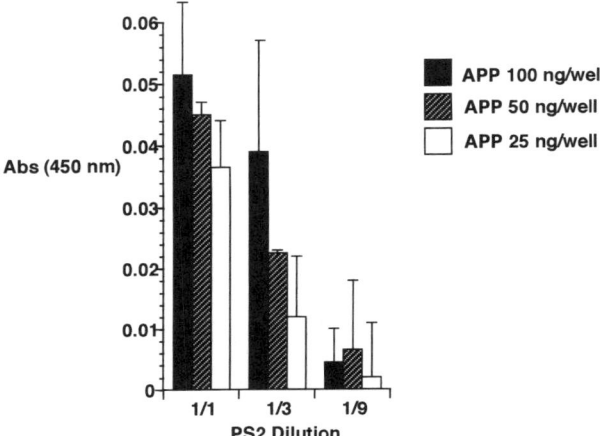

Fig. 4. Detection of APP-PS2 interaction by modified ELISA assay. Three concentrations of purified secreted APP (APPs) were captured on zinc(II) coated microtiter wells and increasing amounts of partially purified epitope tagged PS2 were added to the microtiter wells. The amount of PS2 bound to the microtiter dish was measured using M2 monoclonal antibody specific for the PS2 produced under inducing conditions. PS2 did not bind to wells in which APPs was omitted; neither was any significant binding detected when extracts from uninduced H4 cells were added to the APP-containing wells

Discussion

It is clear that up to 50% of early-onset FAD is caused by mutations in PS1 and PS2 and, although a number of studies have directly linked the mutations in both PS1 and PS2 to increased levels of Aβ, the mechanism by which these mutations cause the observed changes in Aβ production remains unclear. Given that immunohistochemical studies indicate that APP and the presenilins are both found in membranes within similar intracellular compartments (ER/Golgi), that there is a detectable interaction between PS2 and APP, that APP must be properly processed to avoid Aβ and that the *C. elegans* presenilin homologues have been postulated to be involved in cytoplasmic trafficking/processing of proteins, we feel that it is possible that the presenilins play a role in the cytoplasmic partitioning, trafficking and/or transport of APP. The findings presented here indicate that the N141I mutant forms of PS2 are able to be co-immunoprecipitated with APP and suggest that this FAD-associated mutation does not grossly alter the ability of the two proteins to interact. However, the observation that the N141I forms of PS2 are not able to crosslink to APP as efficiently as the wild-type molecule indicates that it is possible that the FAD-associated mutation does subtly alter this interaction and ultimately results in sufficient misprocessing of APP to allow for the observed alterations in Aβ production that are associated with FAD. Perhaps the N141I mutation, which is located at a membrane/lumenal junction, alters the conformation of PS2 within the membrane, and this altered conformation is detected by the in vivo crosslinking paradigm but not in the coimmunoprecipitation paradigm, which necessitates that proteins are extracted from their normal membranous milieu. A second possibility, which we have not explored here, is that the FAD-associated mutations alter the ability of the presenilins to interact with other, as-of-yet unidentified molecules in the Aβ-producing pathway.

Recent results from a number of laboratories indicate that both PS1 and PS2 are endoproteolytically cleaved to produce C- and N-terminal fragments. In addition, it has been demonstrated that under certain conditions an alternative cleavage of PS2 is favored, that the majority of the resulting alternative C-terminal fragment remains associated with the detergent-insoluble cellular fraction and that, when compared to wild-type, this fragment turns over more slowly in cell lines that express the N141I Volga German mutation (Kim et al. 1997; Tanzi et al., this volume). Treatment with proteasome inhibitors leads to increased accumulation of polyubiquinitated PS2 as well as the C-terminal fragment. Thus, it appears possible that PS2 containing the N141I FAD mutation is not efficiently degraded via the proteasome pathway and that this results in a slower turnover and increased accumulation of the C-terminal fragments. The APP-PS2 interaction experiments presented here were carried out with protein extracts prepared from samples in which the majority of PS2 produced was full length (as opposed to the high molecular weight aggregated material or the cleaved fragments), suggesting that APP is able to interact with full length PS2. It is not yet known whether the APP-PS2 interaction is affected by the cleavage of PS2. Future experiments must address this issue as well as the identification of the specific amino acids in both PS2 and APP that are involved in this interaction.

APP has been shown to interact with a number of biological molecules since its discovery in 1987; however, our demonstration of a interaction of APP with PS2, a novel molecule that is believed to be located primarily in the membranes of the ER and Golgi, is the first to be reported for interaction of intracellular APP. Our finding that PS2 is able to interact with secreted APP in the modified ELISA assay indicated that PS2 interacts with the N-terminal portion of APP. Interestingly, the eight transmembrane model of PS2 dictates that the portions of the molecule available to interact with this domain of APP (while the two molecules are in the intracellular membranes of the ER/Golgi) are the smaller loop domains, as opposed to the N- or C-termini of PS2 or the large loop located between TMD 6/7. Thus, the interaction with APP described here might be predicted to take place within the lumenal compartments of the ER/Golgi and would not preclude interaction with proteins in the cytoplasm via the N- and C-terminal and large loop regions of PS2.

Acknowledgments. We would like to acknowledge T.-W. Kim for providing the tetracycline-regulated PS2 H4 neuroglioma cell lines, C. Masters for the APPs, and K. Beyreuther, D. Selkoe and Athena Neurosciences for providing antibodies. We thank S. Guénette, D. Kovacs and T.-W. Kim for helpful discussions. This work was supported by grants from the USPHS grants AG11899 and NS35975 (to WW) and AG11337 (to RET).

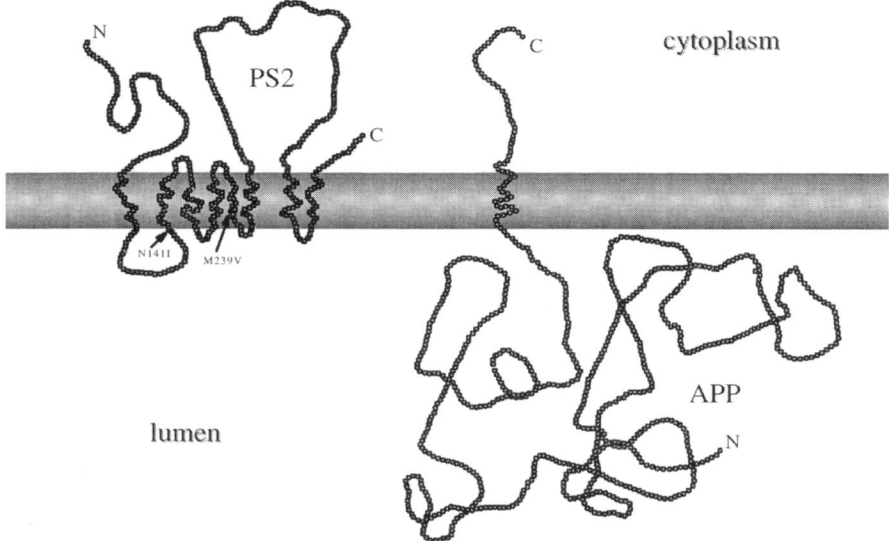

Fig. 5. APP and PS2 in the membranes of the ER/Golgi

References

Borchelt DR, Thinakaran G, Eckman CB, Lee MK, Davenport F, Ratovitsky T, Prada C-M, Kim G, Seekins S, Yager D, Slunt HH, Wang R, Seeger M, Levey AI, Gandy SE, Copeland NG, Jenkins NA, Price DL, Younkin SG, Sisodia SS (1996) Familial Alzheimer's disease-linked presenilin 1 variants elevate Aβ1-42/1-40 ratio in vitro and in vivo. Neuron 17: 1005–1013

Bush AI, Pettingell WH, Paradis MD, Tanzi RE (1994) Modulation of Aβ adhesiveness and secretase site cleavage by zinc. J Biol Chem 269: 26618–26621

Buxbaum JD, Greengard P (1996) Regulation of APP processing by intra- and intercelullar signals. Ann N Y Acad Sci 777: 327–331

Cai X-D, Golde TE, Younkin SG (1993) Release of excess amyloid β protein from mutant amyloid β protein precursor. Science 259: 514–516

Chartier-Harlin M, Crawford F, Houlden H, Warren A, Hughes D, Fidani L, Goate A, Rossor M, Roques P, Hardy J, Mullan M (1991) Early-onset Alzheimer's disease caused by mutations at codon 717 of the β-amyloid precursor protein gene. Nature 353: 844–841

Citron M, Oltersdorf T, Haass C, McConlogue L, Hung AY, Seubert P, Vigo-Pelfrey C, Lieberburg I, Selkoe DJ (1993) Mutation of the β-amyloid precursor protein in familial Alzheimer's disease increases β-protein production. Nature 360: 672–674

Citron M, Westaway D, Xia W, Carlson G, Diehl TS, Levesque G, Johnson-Wood K, Lee M, Seubert P, Davis A, Kholodenko D, Motter R, Sherrington R, Perry B, Yao H, Strome R, Lieberburg I, Rommens J, Kim S, Schenk D, Fraser P, St George-Hyslop P, Selkoe DJ (1997) Mutant presenilins of Alzheimer's disease increase production of 42 residue amyloid β-protein in both transfected cells and transgenic mice. Nature Med 3: 67–72

Corder EH, Saunders AM, Strittmatter WJ, Schmechel DE, Gaskell PC, Small GW, Roses AD, Haines JL, Pericak-Vance MA (1993) Gene dose of apolipoprotein E type 4 allele and the risk of Alzheimer's disease in late onset families. Science 261: 921–923

Deng G, Pike CJ, Cotman CW (1996) Alzheimer-associated presenilin-2 confers increased sensitivity to apoptosis in PC12 cells. FEBS Lett 397: 50–54

Doan A, Thinakaran G, Borchelt DR, Slunt HH, Ratovitsky T, Podlisny M, Selkoe DJ, Seeger M, Gandy SE, Price DL, Sisodia SS (1996) Protein topology of presenilin 1. Neuron 17: 1023–1030

Gandy SE, Greengard P (1994) Processing of Aβ-amyloid precursor protein: Cell biology, regulation and role in Alzheimer's disease. Int Rev Neurobiol 36: 29–50

Goate A, Chartier-Harlin M, Mullan M, Brown J, Crawford F, Fidani L, Giuffra L, Haynes A, Irving N, James L, Mant R, Netwon P, Rooke K, Roques P, Talbot C, Pericak-Vance M, Roses A, Williamson R, Rossor M, Owen M, Hardy J (1991) Segregation of a missense mutation in the amyloid precursor protein gene with familial Alzheimer's disease. Nature 349: 704–706

Gossen M, Bujard H (1992) Tight control of gene expression in mammalian cells by tetracycline-responsive promoters. Proc Natl Acad Sci USA 89: 5547–5551

Hardy J (1997) Amyloid, the presenilins and Alzheimer's disease. Trends Neurosci 20: 154–159

Hendriks L, van Dujin CM, Cras P, Cruts M, van Hul W, van Harskamp F, Martin J-J, Hofman A, van Broeckhoven C (1992) Presenile dementia and cerebral hemorrhage linked to a mutation at codon 692 of the β-amyloid precursor protein. Nature Genet 1: 218–222

Kim T-W, Pettingell WH, Hallmark OG, Moir RD, Wasco W, Tanzi RE (1997) Endoproteolytic cleavage and proteasomal degradation of presenilin 2 in transfected cells. J Biol Chem 272: 11006–11010

Kovacs DM, Fausett HJ, Page KJ, Kim T-W, Mori RD, Merriam DE, Hoillister RD, Hallmark OG, Mancini R, Felsenstein KM, Hyman BT, Tanzi RE, Wasco W (1996) Alzheimer associated presenilins 1 and 2: Neuronal expression in brain and localization to intracellular membranes in mammalian cells. Nature Med 2: 224–229

Lehmann S, Chiesa R, Harris DA (1997) Evidence for a six-transmembrane domain structure of presenilin 1. J Biol Chem 272: 12047–12051

Levitan D, Greenwald I (1995) Facilitation of the *lin-12*-mediated signalling by *sel-12*, a *Caenorhabditis elegans* S182 Alzheimer's disease gene. Nature 377: 351–354

Levy E, Carman MD, Fernandez-Madrid IJ, Power MD, Lieberburg I, Van Duinen SG, Bots GTAM, Luyendijk W, Frangione B (1990) Mutation of the Alzheimer's disease amyloid gene in hereditary cerebral hemorrhage, Dutch type. Science 248: 1124–1126

Levy-Lahad E, Wasco W, Poorkaj P, Romano DM, Oshima J, Pettingell WH, Yu C, Jondro PD, Schmidt SD, Wang K, Crowley AC, Fu Y-H, Guenette SY, Galas D, Nemens E, Wijsman EM, Bird TD, Schellenberg GD, Tanzi RE (1995) Candidate gene for the chromosome 1 familial Alzheimer's disease locus. Science 269: 973–977

L'Hernault SW, Arduengo PM (1992) Mutation of a putative sperm membrane protein in *Caenorhabditis elegans* prevents sperm differentiation but not is associated meiotic divisions. J Cell Biol 119: 55–68

Li X, Greenwald I (1996) Membrane topology of the C. elegans SEL-12 presenilin. Neuron 17: 1015–1021

Mellman L, Matter K, Yamamoto E, Pollack N, Roome J, Felsenstein K, Roberts S (1995) Mechanisms of molecular sorting in polarized cells: relevance to Alzheimer's disease. In: Kosik KS, Christen Y, Selkoe DJ (eds) Alzheimer's disease: lessons from cell biology. Springer-Verlag, pp 14–26

Mullan M, Crawford F, Axelman K, Houlden H, Lilius L, Winblad W, Lannfelt L (1992) A pathogenic mutation of probable Alzheimer's disease in the N-terminus of β-amyloid. Nature Genet 1: 345–347

Murrell J, Farlow M, Ghetti B, Benson MD (1991) A mutation in the amyloid precursor protein associated with hereditary Alzheimer's disease. Science 254: 97–99

Perez-Tur J, Froelich S, Pirhar G, Crook R, Baker M, Duff K, Wragg M, Busfield F, Lendon C, Clark RF, Roques P, Fuldner RA, Hohnston J, Cowburn R, Forsell C, Axelman K, Lilius L, Houlden H, Karren E, Roberts GW, Rossor M, Adams MD, Hardy J, Goate A, Lannfelt L, Hutton M (1996) A mutation in Alzheimer's disease destroying a splice acceptor site in the presenilin 1 gene. Neuroreport 7: 297–301

Pericak-Vance MA, Bebout JL, Gaskell PC Jr, Yamaoka LH, Hung WY, Alberts MJ, Walker AP, Bartlett RJ, Haynes CA, Welsh KA, Earl NL, Heyman A, Clark CM, Roses AD (1991) Linkage studies in familial Alzheimer disease: evidence for chromosome 19 linkage. Am J Human Genet 48: 1034–1050

Rebeck GW, Reiter JS, Strickland DK, Hyman BT (1993) Apolipoprotein E in sporadic Alzheimer's disease: allelic variation and receptor interactions. Neuron 11: 575–580

Rogaev EI, Sherrington R, Rogaeva EA, Levesque GM, Liang Y, Chi H, Lin C, Holman K, Tsuda T, Mar L, Sorbi S, Nacmias B, Piacentini S, Amaducci L, Chumakov I, Cohen D, Lannfelt L, Fraser PE, Rommens JM, St George-Hyslop PH (1995) Familial Alzheimer's disease in kindreds with missense mutations in a gene on chromosome 1 related to the Alzheimer's disease type 3 gene. Nature 376: 775–778

Selkoe DJ (1995) Physiological production and polarized secretion of the amyloid β-peptide in endothelial cells: A route to the mechanism of Alzheimer's disease. In Kosik KS, Christen Y, Selkoe DJ (eds) Alzheimer's disease: lessons from cell biology. Berlin, Springer-Verlag, pp 70–77

Scheuner D, Eckman C, Jensen M, Song X, Citron M, Suzuki N, Bird TD, Hardy J, Hutton M, Kukull W, Larson E, Levy-Lahad E, Viitanen M, Peskind E, Poorkaj P, Schellenberg G, Tanzi RE, Wasco W, Lannfelt L, Selkoe D, Younkin S (1996) Secreted amyloid β-protein similar to that in senile plaques of Alzheimer's disease is increased in vivo by the PS1/2 and APP mutations linked to familial Alzheimer's disease. Nature Med 2: 865–870

Sherrington R, Rogaev EI, Liang Y, Rogaeva EA, Levesque G, Ikeda M, Chi H, Lin C, Li G, Holman K, Tsuda T, Mar L, Foncin J-F, Bruni AC, Montesi M, Sorbi S, Rainero I, Pinessi L, Nee L, Chumakov Y, Pollen D, Wasco W, Haines JL, DaSilva R, Pericak-Vance M, Roses AD, Tanzi RE, Fraser PE, Fraser P, Suzuki N, Cheung TT, Cai XD, Odaka A, Otvos L Jr, Eckman C, Golde TE, Rommens JM, St George-Hyslop PH (1995) Cloning of a novel gene bearing missence mutations in early onset familial Alzheimer's disease. Nature 375: 754–760

Strittmatter WJ, Saunders AM, Schmechel D, Pericak-Vance M, Enghild J, Salvesen GS, Roses AD (1993) Apolipoprotein E: high avidity binding to β-amyloid and increased frequency of type 4 allele in late-onset familial Alzheimer disease. Proc Natl Acad Sci USA 90: 1977–1981

Suzuki N, Cheung TT, Cai XD, Odaka A, Otvos L Jr, Eckman C, Golde TE, Younkin SG (1994) An increased percentage of long amyloid β protein secreted by familial amyloid β protein precursor (β APP717) mutants. Science 264: 1336–1340

Thinakaran G, Borchelt D, Lee M, Slunt H, Spitzer L, Kim G, Ratovitsky T, Davenport F, Nordstedt C, Seeger M, Hardy J, Levey AI, Gandy SE, Jenkins NA, Copeland NG, Price DL, Sisodia SS (1996) Endoproteolysis of Presenilin 1 and accumulation of processed derivatives in vivo. Neuron 17: 181–190

Vito P, Lancana E, D'Adamio L (1996) Interfering with apoptosis: Ca^{2+}-binding protein ALG-2 and Alzheimer's disease gene ALG-3. Science 271: 521–525

Wolozin B, Iwasaki K, Vito P, Kelly G, Lacana E, Sutherland T, Zhao B, Kusiak J, Wasco W, D'Adamio L (1996) PS2 participates in cellular apoptosis: Constitutive activity conferred by Alzheimer mutation. Science 274: 1710–1713

Wasco W, Tanzi RE (1996) Etiological clues from gene defects causing early-onset familial Alzheimer's disease. In: Wasco W, Tanzi RE, Torowa NS (eds) Molecular models of dementia Toronto. The Humana Press Inc, pp 1–20

The Cellular Biology of Presenilin Proteins and a Novel Mechanism of Amyloid β-Peptide Generation

C. Haass[1], J. Walter[1], A. Capell[1], C. Wild-Bode[1], J. Grünberg[1], T. Yamazaki[2], I. Ihara[2], I. Zweckbronner[3], C. Jakubek[3], and R. Baumeister[3]

Presenilin (PS) proteins are involved in numerous cases of familial Alzheimer's disease (reviewed by Tanzi et al. 1996) and are therefore key players in the pathogenesis of the disease. To understand the biology of the PS proteins we searched for a potential biological function and analyzed the biochemistry of the two homologous PS proteins. We also determined the subcellular localization of presenilins and took advantage of this knowledge to search for a cellular mechanism that is involved in Aβ42 generation. These three main topics will be reviewed in detail.

A Biological Function of Presenilin Proteins in Notch Signaling

Cloning of the presenilin (PS) homologous *sel-12* gene in *Caenorhabditis elegans* (Levitan and Greenwald 1995) opened up the possibility that human PS genes might be directly or indirectly involved in Notch signaling. To prove this possibility, we expressed human PS minigenes under the control of the *sel-12* promotor in *sel-12* mutant worms (Baumeister et al. 1997). Expression of wt human PS1 in mutant animals rescued all aspects of the mutant phenotype. Animals accumulated the normal number of eggs in their bodies, they were of normal size, and they behaved normally (Table 1). These findings clearly indicate that human PS1 can functionally replace defective SEL-12 in *C. elegans*, making it likely that PS proteins might be involved in Notch signaling either directly or indirectly.

We then repeated these experiments but expressed two different PS1 point mutations (A246E, C410Y) that are known to cause familial Alzheimer's disease (FAD) in humans (reviewed by Tanzi et al. 1996). Surprisingly, these mutations exhibited a dramatically reduced ability to rescue the mutant *sel-12* phenotype (Table 1). The failure of mutant PS molecules to rescue the mutant phenotype in *C. elegans* indicates that the mutations occurred at functionally important posi-

---

[1] Central Institute for Mental Health, Department of Molecular Biology, J5, 68159 Mannheim, GERMANY;
[2] Department of Neuropathology, Inst. for Brain Research, Faculty of Medicine, University of Tokyo, Tokyo 113, JAPAN;
[3] Laboratory of Molecular Biology/Genzentrum of the University of Munich, 81377 Munich, GERMANY.

S. G. Younkin / R. E. Tanzi / Y. Christen (Eds.)
Presenilins and Alzheimer's Disease
© Springer-Verlag Berlin Heidelberg New York 1998

Table 1. Rescue of the *sel-12* mutant phenotype by the indicated cDNA constructs

cDNA construct	Rescue of the *sel-12* mutant phenotype
sel-12	++++
PS1 wt	++++
PS1 A246E	–
PS1 C410Y	–
PS1 TM7 STOP	++++
PS1 TM6 STOP	–
PS1 TM5 STOP	–
PS1 TM4 STOP	–
PS1 ΔExon 10	+++

tions of the molecule. This finding is strongly supported by the fact that all FAD-associated mutations occur at positions that are conserved between *C. elegans* and human presenilins. It therefore appears that FAD-associated mutations could interfere with the biological activity of presenilins by exchanging single amino acids located at functionally and for structurally important positions.

Having a biological assay in hand, we mapped a biologically active domain of PS1 by a deletion analysis. Stop codons were inserted after the transmembrane (TM) domains 7, 6, 5 and 4 and the truncated proteins expressed under the control of the *sel-12* promotor. We found that the cytoplasmic tail C-terminal of TM7 is dispensable for the biological activity of PS1 in mutant worms. However, insertion of a stop codon after TM6, which results in the additional deletion of the large cytoplasmic loop, results in a severe loss of biological activity (Table 1). Therefore the large loop is of functional importance for the PS proteins. One should note, however, that the above-mentioned FAD-associated point mutations occur outside of the large loop and also affect the biological function of PS proteins. We therefore find it very likely that structural changes, even those induced by a single amino acid exchange, can be sufficient to cause a loss of biological function of presenilins.

There is currently heavy discussion in the field regarding whether proteolytic processing (summarized by Tanzi et al. 1996) is required for the biological activity of PS proteins. This discussion is based on the finding that very little full-length PS can be observed *in vivo*, whereas abundant proteolytic fragments of ~30 kDa (N-terminal fragment, NTF) and ~20 kDa (C-terminal fragment, CTF) are detected (Thinakaran et al. 1996). To prove this, we took advantage of the finding by Thinakaran et al. (1996) that a FAD-causing splicing mutation (Δexon 10) that removes the domain encoded by exon 10 inhibits proteolytic processing of PS1. Surprisingly, expression of the Δexon 10 cDNA in the mutant worm rescued the mutant *sel-12* phenotype almost perfectly (Table 1). In parallel we showed that wt human PS1 is proteolytically processed in transgenic worms like in human cells and tissues, whereas PS1 Δexon 10 does not undergo proteolytic processing. This finding clearly indicates that proteolytic processing of PS1 is not an absolute prerequisite for the biological activity of PS1 in *C. elegans*.

However, this also raises another interesting problem. The Δexon 10 mutation is a FAD-associated mutation. As shown above, FAD-associated point mutations clearly lose their biological activity, whereas the splicing mutation retains at least some biological activity. One therefore has to argue that point mutations affect biological mechanisms other than the Δexon 10 mutation.

Finally, we also analyzed the cellular expression of PS1 in *C. elegans*. We expressed the Green Fluorescent Protein (GFP) and PS1 under the control of the endogenous *sel-12* promotor. Tissue and cell specific expression was then determined by analyzing the autofluorescence of GFP or by immunocytochemistry using our previously established antibodies to a variety of PS1 domains (Walter et al. 1996). We found that PS proteins were expressed not only in gonads but also in the nervous system and some muscle cells (Fig. 1). These results clearly indicate that *SEL-12* plays an important functional role in neurons, the cell type that is most severely affected by AD.

Taken together, our results (and those independently obtained by Levitan et al. 1996) strongly indicate that PS proteins are indeed involved directly or indirectly in Notch signaling and might have an important function in nerve cells. By

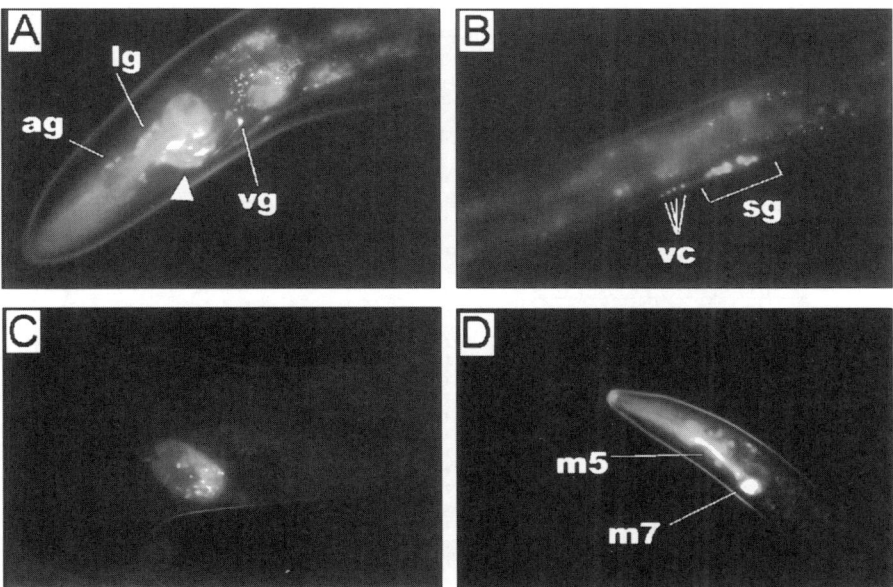

Fig. 1. Expression of *sel-12* in *C. elegans*. (A) Nerve ring neuropile (arrowhead), cell body and neurites of interneuron AVK in the ventral ganglion (vg), lateral (lg), and anterior (ag) ganglion neurons. (B) *sel-12* expression in the somatic gonad (sg) and ventral cord neurons (vc). (C) *sel-12* expression in a single embryo in the uterus of a transgenic animal. (D) *sel-12* expression in the pharyngeal muscles m5 and m7. *sel-12* expression is visualized by using a sel-12::GFP construct. Identical results were obtained after expression of PS1 under the control of the *sel-12* promotor followed by immunostaining using PS1 specific polyclonal antibodies (data not shown)

an unknown mechanism FAD-associated point mutations appear to interfere with these biological functions.

Based on data presented by Levitan et al. (1996), PS1 as well as PS2 can rescue the mutant *sel-12* phenotype, which indicates that both proteins might be functionally redundant. However, we (Walter et al. 1996, 1997) and others (Seeger et al. 1997) found that both presenilins are biochemically different in terms of posttranslational modifications, indicating differential functional activities.

Differential Phosphorylation of PS Proteins

We carried out a analysis of PS phosphorylation and found that PS proteins are differentially phosphorylated in very complex manner specific to PS1 and PS2 (Walter et al. 1996, 1997; Walter and Haass, manuscript in preparation).

Constitutive Phosphorylation of PS2

PS2 is constitutively phosphorylated as a full-length molecule. Three major phosphorylation sites were detected within an acidic domain missing in PS1 (Fig. 2). Phosphorylation occurs at serine residues 7, 9 and 19 most likely by Casein kinase (CK) I and CKII (Walter et al. 1996). In addition to the N-terminal phos-

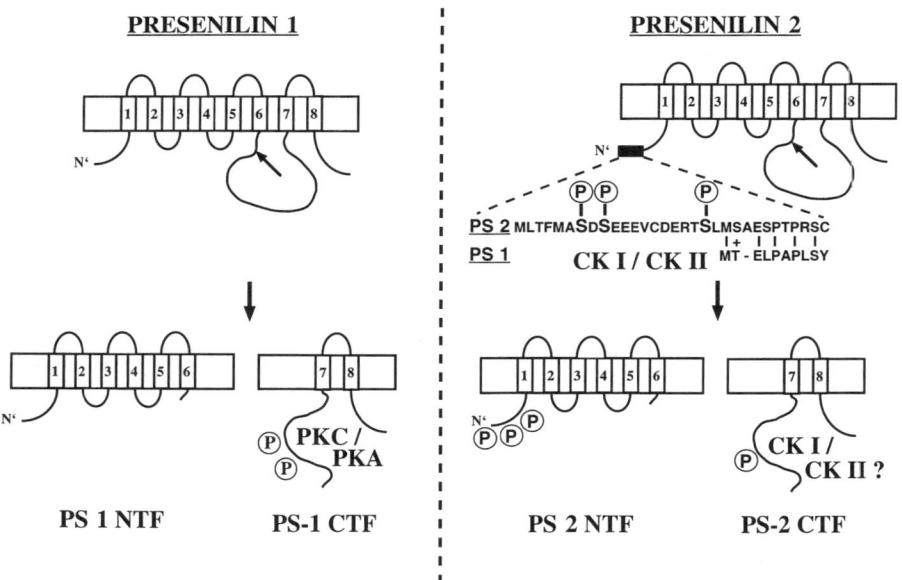

Fig. 2. Schematic representation of the differential phosphorylation of PS1 and PS2

phorylation, we recently also detected a phosphorylation site within the CTF of PS2 where again a serine appears to be phosphorylated by a CK-like kinase (Walter and Haass, manuscript in preparation). Interestingly, we found that full-length PS1 is phosphorylated very little if at all, indicating a major biochemical difference between PS1 and PS2.

Protein kinase C and protein kinase A-dependent phosphorylation of PS1

Upon stimulation of protein kinase C with phorbolester, we obtained an increased molecular weight of the PS1 $\sim$20 kDa CTF which was shifted to $\sim$23 kDa (Seeger et al. 1997, Walter et al. 1997). It appears that phosphorylation of the PS1 CTF results in the characteristic molecular weight shift, since treatment with alkaline phosphatase reverses this phenomenon. In addition to PKC, we found that PKA can also phosphorylate within the large loop between TM6 and TM7. Phosphoamino acid analysis revealed that the PS1 CTF is phosphorylated on serines. Interestingly, we have observed PKC- and PKA-dependent phosphorylation only after proteolytic processing of PS1, whereas so far no phosphorylated full-length PS1 could be detected. This leads to the hypothesis that proteolytic cleavage of PS1 unfolds the corresponding phosphorylation sites, making them accessible for PKC and PKA (Fig. 2). Very recently we found that PKC/PKA phosphorylation is PS1 specific and does not occur on the corresponding PS2 CTF (Walter and Haass, manuscript in preparation).

So far nothing is known about the biological function of the differential phosphorylation of presenilins; however, in the case of PS1, one might argue that the regulated phosphorylation could be involved in CTF turnover. Another possibility would be that CTF phosphorylation determines its subcellular localization.

A Novel Cellular Mechanism of Amyloid β-Peptide (Aβ1-42) Generation

Recently several groups reported that mutations within the PS genes increase the production of the 42 amino acid version of Aβ (Aβ42; Scheuner et al. 1995; Borchelt et al. 1996; Duff et al. 1996; Tomita et al. 1996; Xia et al. 1997; Citron et al. 1997). This result is very surprising, since PS proteins are predominantly located within the endoplasmic reticulum (ER; Kovacs et al. 1996; Cook et al. 1996; Walter et al. 1996; DeStrooper et al. 1997) whereas βAPP is predominantly detected within the Golgi and on the cell surface (Selkoe 1996). Moreover, Aβ generation was shown to occur within endosomes and at or close to the cell surface (Selkoe 1996). It is therefore very difficult to explain the effect of PS mutations on Aβ generation, since PS proteins and the machinery involved in Aβ generation are located in different cellular compartments. We hypothesized that Aβ42 might be generated in a compartment that is different from that where Aβ40 (which represents more than 90% of secreted Aβ) is known to be cleaved (Wild-Bode et al.,

1997). A first indication for the presence of such a novel pathway would be the detection of intracellular Aβ42. Indeed, we were able to detect intracellular Aβ42 by using a previously established ELISA. Moreover, we could show that the FAD-associated βAPP V717G mutation results in an increased level of intracellular Aβ42 similar to that reported previously for secreted Aβ42 (Suzuki et al. 1994). Interestingly, we detected more intracellular Aβ42 as Aβ40, which is the opposite of what is found in conditioned media. This indicates that Aβ42 might accumulate intracellularly, probably due to its ability to aggregate, whereas the soluble Aβ40 is secreted more efficiently. Since PS proteins are predominantly located within the endoplasmic reticulum (ER) we retained βAPP in the ER by treatment with brefeldin A (BFA) and determined the levels of secreted and intracellular Aβ42. As expected BFA treatment caused the inhibition of secretion of both species of Aβ. However, within cell lysates BFA caused an increased production of Aβ42. This finding might indicate that the ER is at least one of the sites of Aβ42 generation (Table 2). However, treatment of cells with monensin, which accumulates βAPP within the Golgi, also enhanced Aβ42 production, indicating that at least some Aβ42 molecules can be generated after the release of the precursor from the ER. Endosomal processing is not involved in Aβ42 generation since inhibition of endosomal proteases by NH_4Cl and inhibition of reinternalization did not inhibit Aβ42 production (Table 2). Therefore it appears that Aβ42 is preferentially generated within the same compartement(s) where the PS proteins are located. Thus, these results raise the possibility that PS proteins might directly interact with βAPP within the ER in a manner that prevents (wt PS) Aβ42 generation or allows (mutant PS) Aβ42 generation. Such a possibility is strongly supported by the recent findings that PS proteins can indeed bind to immature βAPP (Weidemann et al. 1997).

Acknowledgment. This work was support by the Boehringer Ingelheim KG and grants from the Deutsche Forschungsgemeinschaft (SFB 317) and the Thyssen Foundation (to CH).

Table 2. Cellular localization of secretases

Compartment	β-secretase	γ-secretase (40)	γ-secretase (42)
Endoplasmic Reticulum	+	−	++
Golgi	++	−	+
Cell surface	?	+++	++
Endosomes	+++	+++	−

References

Baumeister R, Leimer U, Zweckbronner I, Jakubek C, Grünberg J, Haass C (1997) Human presenilin-1, but not familial Alzheimer's disease (FAD) mutants, facilitate *Caenorhabditis elegans* Noth signalling independently of proteolytic processing. Genes Function, 1: 149–159

Borchelt DR, Thinakaran G, Eckman CB, Lee MK, Davenport F, Ratovitsky T, Prada C-M, Kim G, Seekins S, Yager D, Slunt HH, Wang R, Seeger M, Levey AI, Gandy SE, Copeland NG, Jenkins NA, Price DL, Younkin SG, Sisodia SS (1996) Familial Alzheimer's disease-linkes presenilin 1 variants elevate Aβ1-42/1-40 ratio in vitro and in vivo. Neuron 17: 1005–1013

Citron M, Westaway D, Xia W, Carlson G, Diehl T, Levesque G, Johnson-Wood K, Lee M, Seubert P, Davis A, Kholodenko D, Motter R, Sherrington R, Perry B, Yao H, Strome R, Lieberburg I, Rommens J, Kim S, Schenk D, Fraser P, St George Hyslop P, Selkoe DJ (1997) Mutant presenilins of Alzheimer's disease increase production of 42-residue amyloid β-protein in both transfected cells and transgenic mice. Nature Med 3: 67–72

Cook DG, Sung JC, Golde TE, Felsenstein KM, Wojczek BS, Tanzi RE, Trojanowski JQ, Lee V M-Y, Doms RW (1996) Expression and analysis of presenilin 1 in a human neuronal system: localization in cell bodies and dendrites. Proc Natl Acad Sci USA 93: 9223–9228

De Strooper B, Beullens M, Contreras B, Craessaerts K, Moechars D, Bollen M, Fraser P, St George-Hyslop P, Van Leuven F (1997) Postranslational modification, subcellular localization and membrane orientation of the Alzheimer's disease associated presenilins. J Biol Chem 272: 3590–3598

Duff K, Eckman C, Zehr C, Yu X, Prada C-M, Perez-Tur J, Hutton M, Buee L, Harigaya Y, Yager D, Morgan D, Gordon MN, Holcomb L, Refolo L, Zenk B, Hardy J, Younkin S (1996) Increased amyloid-β42(43) in brains of mice expressing mutant presinilin 1. Nature 383: 710–713

Kovacs DM, Fausett HJ, Page KJ, Kim T-W, Mori RD, Merriam DE, Hoillister RD, Hallmark OG, Mancini R, Felsenstein KM, Hyman BT, Tanzi RE, Wasco W (1996) Alzheimer associated presenilins 1 and 2: neuronal expression in brain and localization to intracellular membranes in mammalian cells. Nat Med 2: 224–229

Levitan D, Greenwald I (1995) Facilitation of *lin-12*-mediated signalling by *sel-12*, a *Caenorhabditis elegans* S182 Alzheimer's disease gene. Nature 377: 351–354

Levitan D, Doyle TG, Brousseau D, Lee MK, Thinakaran G, Slunt HH, Sisodia SS, Greenwald I (1996) Assessment of normal and mutant human presenilin function in *Caenorhabditis elegans*. Proc Natl Acad Sci USA 93: 14940–14944

Scheuner D, Eckman C, Jensen M, Song X, Citron M, Suzuki N, Bird TD, Hardy J, Hutton M, Kukull W, Larson E, Levy-Lahad E, Viitanen M, Peskind E, Poorkaj P, Schellenberg G, Tanzi R, Wasco W, Lannfelt L, Selkoe D, Younkin S (1996) Secreted amyloid β-protein similar to that in the senile plaques of Alzheimer's disease is increased *in vivo* by the presenilin 1 and 2 and *APP* mutations linked to familial Alzheimer's disease. Nature Med 2: 864–852

Seeger M, Nordstedt C, Petanceska S, Kovacs DM, Gouras GK, Hahne S, Fraser P, Levesque L, Czernik AJ, St George-Hyslop P, Sisodia SS, Thinakaran G, Tanzi RE, Greengard P, Gandy S (1997) Evidence for phosphorylation and oligomeric assembly of presenilin 1. Proc Natl Acad Sci USA 94: 5090–5094

Selkoe DJ (1996) Amyloid β-protein and the genetics of Alzheimer's disease. J Biol Chem 271: 18295–18298

Suzuki N, Cheung TT, Cai XD, Odaka A, Otvos L Jr, Eckman C, Golde TE, Younkin SG (1994) An increased percentage of long amyloid β protein secreted by familial amyloid β protein precursor (β APP717) mutants. Science 264: 1336–1340

Tanzi RE, Kovacs DM, Kim T-W, Moir RD, Guenette SY, Wasco W (1996a) The gene defects responsible for familial Alzheimer's disease. Neurobiol Disease 3: 159–168

Thinakaran G, Borchelt D, Lee M, Slunt H, Spitzer L, Kim G, Ratovitsky T, Davenport F, Nordstedt C, Seeger M, Hardy J, Levey AI, Gandy SE, Jenkins NA, Copeland NG, Price DL, Sisodia SS (1996) Endoproteolysis of presenilin 1 and accumulation of processed derivatives in vivo. Neuron 17: 181–190

Tomita T, Maruyama K, Saido TC, Kume H, Shinozaki K, Tokuhiro S, Capell A, Walter J, Gruenberg J, Haass C, Iwatsubo T, Obata K (1997) The presenilin 2 mutation (N141I) linked to familial Alzheimer disease (Volga German families) increases the secretion of amyloid β protein ending at the 42nd (or 43rd) residue. Proc Natl. Acad Sci USA 94: 2025–2030

Walter J, Capell A, Grunberg J, Pesold B, Schindzielorz A, Prior R, Podlisny MB, Fraser P, St George Hyslop P, Selkoe DJ, Haass C (1996) The Alzheimer's disease-associated presenilins are differentially phosphorylated proteins located predominantly within the endoplasmic reticulum. Mol Med 2: 673–691

Walter J, Grünberg J, Capell A, Peshold B, Schindzielorz A, Citron M, Mendla K, St George-Gyslop P, Mutlhaup G, Selkoe DJ, Haass C (1997) Proteolytic processing of the Alzheimer's disease-associated presenilin-1 generates an *in vivo* substrate for protein kinase C. Proc Natl Acad Sci USA **94**: 5349–5354

Weidemann A, Paliga K, Dürrwang U, Czech C, Evin G, Masters CL, Beyreuther K (1997) Formation of stable complexes between two Alzheimer's disease gene products: presenilin-2 and β-amyloid precursor protein. Nature Med **3**: 328–332

Wild-Bode C, Yamazaki T., Capell A, Leimer U, Steiner H, Ihara Y, Haass C (1997) Intracellular generation and accumulation of amyloid β-peptide terminating at aminoacid 42. J Biol Chem 272: 16085–16088

Xia WM, Zhang JM, Kholodenko D, Citron M, Podlisny MB, Teplow DB, Haass C, Seubert P, Koo EH, Selkoe DJ (1997) Enhanced production and oligomerization of the 42-residue amyloid β-protein by Chinese hamster ovary cells stably expressing mutant presenilins. J Biol Chem 272: 7977–7982

Regulation of Presenilin 1 Phosphorylation and Transcriptional Activation of Signal Transduction-Induced Genes by Muscarinic Receptors

U. Langer, C. Albrecht, M. Mayhaus, J. Velden, H. Wiegmann, J. Klaudiny,
D. Müller, H. von der Kammer, and R. M. Nitsch[*]

Summary

To identify neurotransmitter- and activity-dependent post-translational modifications of presenilin 1 (PS), we overexpressed muscarinic acetylcholine receptor subtypes in 293 cell lines and stimulated them with carbachol. We found that the carboxyl-terminal fragment (CTF) of PS1 was readily phosphorylated by the protein kinase (PKC)-coupled subtypes m1 and m3, but not by m2 or m4 receptors. The surface receptor-mediated phosphorylation was mediated by PKC, and it resulted in a shift in the apparent molecular mass of the PS1 CTF in SDS gels. Together with our previous findings, these data demonstrate that the metabolism of both APP and PS1 can be under the control of muscarinic m1 and m3 receptor subtypes. The results suggest the possibility that decreases in neurotransmission coupled to PKC in Alzheimer's disease (AD) brains are associated with deafferentation-dependent abnormalities in the post-translational processing of both APP and PS1.

To test whether the phosphorylation of PS1 derivatives can be involved in APP processing, we overexpressed PS1 and stimulated PKC activity with phorbol esters. We found that overexpression of wild-type PS1, but not PS2 or AD-causing PS1 mutants, accelerated regulated α-secretase processing of APP along with the secretion of APPs. Thus, the increases in $A\beta_{1-42}$ associated with presenilin mutations may be related to changes in regulated α-secretase processing.

By using a differential display screening approach, we found that several transcription factors, including the family of the zinc finger domain early growth response genes EGR-1, EGR-2, EGR-3, EGR-4, are transcriptionally regulated by muscarinic m1 acetylcholine receptors. In addition, we also identified several novel genes – including Gig1 and Gig2 – that are readily expressed as a result of m1 receptor stimulation. The identification and characterization of neurotransmitter-induced genes will help our understanding of deafferentation-induced changes in post-synaptic neurons, and it will help guide the clinical applications of neurotransmitter replacement treatment strategies for Alzheimer's disease.

---

[*] Center for Molecular Neurobiology and Alzheimer's Disease Research Group, University of Hamburg, Germany

Introduction

The neuropathology of AD is characterized by accumulation of extracellular amyloid plaques, cerebrovascular amyloid and intracellular neurofibrillary tangles, along with the degeneration, shrinkage and death of neurons. Primarily, selected population of large projection neurons involved in attention, learning and memory appear to be affected by the neurodegenerative processes. Within the entorhinal cortex, for example, the major projection layers that form the perforant path connection between the hippocampus and the cortex as well as projections to the limbic system are heavily affected by the degenerative process. In affected regions, more than 50% of the neurons are lost during the disease process (Gomez-Isla et al. 1996, 1997). In addition, the large subcortical cholinergic projection neurons that connect the basal forebrain with cortical and hippocampal target cells are lost. The degeneration of projection neurons is accompanied by the loss of synapses and by the decrease in neurotransmitters in their terminal zones. As a result, brain tissue levels of several major neurotransmitters, including acetylcholine, serotonin and glutamate, are reduced in AD (Bowen et al. 1983; Sims et al. 1983; Palmer et al. 1987; Procter et al. 1988; Francis et al. 1993). Together, these observations indicate that post-synaptic target cells in the AD brain receive less than normal input, and they suggest deafferentation of a subset of target neurons involved in attention, learning and memory.

We showed in previous studies that APP processing can be regulated in an activity-dependent fashion by G protein-coupled neurotransmitter receptors, including those for acetylcholine, serotonin and glutamate (Nitsch et al. 1992, 1993, 1996, 1997). In particular, α-secretase processing is readily stimulated by surface receptor activation, possibly at the expense of β-secretase processing, and with the consequence of decreased generation of Aβ peptides (Hung et al. 1993). Increased brain amyloid formation may therefore be related to deafferentation and to pathologically altered activity-dependent processing in post-synaptic neurons that receive less than normal input (for review, see Nitsch and Growdon 1994).

Regulation of Presenilin Phosphorylation by Muscarinic Receptor Subtypes Coupled to Protein Kinase C

Presenilin 1 (PS1; Sherrington et al. 1995) is readily cleaved within its hydrophilic loop domain to generate a 30 kDa N-terminal derivative (NTF) and a 21 kDa C-terminal fragment (CTF) that contains most of the cytoplasmic loop domain (Thinakaran et al. 1996). Both cleavage products contain several putative phosphorylation sites. We found that the PS1 CTF is a substrate for regulated phosphorylation induced by the muscarinic acetylcholine receptor subtypes m1 and m3, but not m2 or m4 receptor subtypes (Langer et al., submitted for publication). Direct activation of PKC by phorbol esters also increases PS1 loop phosphorylation (Seeger et al. 1997; Walter et al. 1997). In addition, activation of PKC also increased cellular levels of PS1 holoprotein without increasing message levels or the amount of the PS1 NTF. These data show that PS1 metabolism can be controlled by cell-surface receptors coupled to PKC, and they suggest activity-dependent regulation of post-translational PS1 processing. Together with our previous data on regulated APP processing, these findings suggest the possibility that surface receptor-dependent PS1 loop phosphorylation is involved in the regulation of secretory APP processing. Together the data imply that the activity-dependent post-translational processing of two proteins – APP and PS1 – with critical roles in the pathogenesis of AD can be under the control of external signals that are substantially decreased in AD brains.

Presenilin Mutation Attenuate Regulated Secretory APP Processing

Many FAD-causing PS mutations are associated with increased generation of $A\beta_{1-42/43}$. (Scheuner et al. 1996). Thus, presenilins may interact with post-translational processing of the amyloid β-protein precursor (APP), and a putative biological function of PS proteins may be related to the control of cellular APP metabolism. PS proteins are predominantly localized in the perinuclear endoplasmic reticulum and the cis-Golgi, where possible interactions with APP could occur (Kovacs et al. 1996). To investigate whether presenilins can interact with regulated α-secretase processing of APP, we stably transfected 293 cells with wild-type presenilins or disease-causing mutants and stimulated α-secretase processing with phorbol myristate acetate (PMA). Stable overexpression of wild-type PS1 accelerated PMA-stimulated, but not constitutive, α-secretase processing of APP several fold, and disease-causing PS mutants M146L, H163R, A246E and L286V failed to do so at similar expression levels. These data indicate that overexpression of wild-type PS1 facilitates regulated α-secretase processing, and they demonstrate a loss of this particular function of FAD-causing mutants. Thus, the reported increases in $A\beta_{1-42}/A\beta_{1-40}$ ratios in patients with PS1 mutations may be associated with lower than normal rates of α-secretase processing.

Activity-Dependent Regulation of Gene Expression by Muscarinic Receptors

To identify activity-dependent and receptor-mediated effects on gene expression in postsynaptic cells, we used a differential screen of cDNA populations (Liang and Pardee 1992) obtained from carbachol-stimulated cells that express m1 receptors and compared them to unstimulated controls. We identified several differentially regulated genes, including the family of early growth response genes EGR-1, EGR-2, EGR-3, and EGR-4 that encode zinc-finger domain transcription factors (Suggs et al. 1990). In addition, we identified Gig1, a previously unknown human homolog of the mouse secretory extracellular matrix signaling protein Cyr61 (O'Brien et al. 1990). Sequence analysis of a 2021 bp full length Gig1 cDNA-clone revealed an open reading frame that encodes a 381 residue polypeptide with 91% homology to the mouse polypeptide sequence. Egr and Gig1 mRNA levels were undetectable by Northern blot analysis in unstimulated cells. Within 10 minutes of m1 receptor stimulation with carbachol, Egr and Gig1 mRNA increased dramatically and attained a maximum at 50 minutes. Mobility shift assays along with Western blots demonstrated synthesis of Egr-1 protein with the ability to bind its DNA recognition sequence as a result of m1-receptor activation. The increase in Egr-1 and Gig1 synthesis was blocked by atropine, mimicked by PMA, and was blunted by downregulation of PKC. By using an Egr-1-specific luciferase reporter assay (Sock et al. 1997), we demonstrated that stimulated m1-receptors activate transcription from an Egr-1-dependent promoter. These findings show that EGR and Gig1 are downstream targets of muscarinic m1 receptor-initiated signaling, and they imply that EGR-dependent target genes are also transcriptionally activated by m1 receptors. Thus, it is likely that muscarinic transmission is associated with both short-term and long-term cellular responses coupled to receptor activation by the transcriptional activation of EGR transcription factors. Future studies will address whether EGR-dependent transcriptional regulation is altered in deafferented target neurons in AD brains, and whether subtype selective ligands are useful in the modulation of EGR-dependent transcription. In addition to transcription factors, the synthesis and secretion of Gig1 protein could also be associated with transmitter-induced long-term alterations in cellular responses, perhaps by growth modulation via possible interactions of Gig1 with extracellular matrix proteins.

Conclusions

Many genes and proteins are regulated in an activity-dependent manner by cell surface receptors. Transcription factors that are activated by muscarinic receptors include immediate early genes, transcription factors, ion channels, several unknown genes as well as the AD-associated proteins APP and PS1. It is therefore possible that reductions in neuronal activity along with the loss of synapses in AD cause abnormal regulation of activity-dependent genes, as well as altered processing and function of APP and PS1. Treatments designed to restore neurotransmission in AD brain may thus be useful for modulating the expression, post-translational modification and cellular function of downstream target molecules coupled to cell surface receptors.

References

Bowen DM, Allen SJ, Benton JS, Goodhardt MJ, Haan EA, Palmer AM, Sims NR, Smith CCT, Spillane JA, Esiri MM, Neary JS, Snowdon JS, Wilcock GK, Davison AN (1983) Biochemical assessment of serotonergic and cholinergic dysfunction and cerebral atrophy in Alzheimer's disease. J Neurochem 41: 266-272

Francis PT, Sims NR, Procter AW, Bowen DM (1993) Cortical pyramidal neurone loss may cause glutamatergic hypoactivity and cognitive impairment in Alzheimer's disease: investigative and therapeutic perspectives. J Neurochem 60: 1589-1604

Gomez-Isla T, Price JL, McKeel DW, Morris JC, Growdon JH, Hyman BT (1996) Profound loss of layer II of entorhinal cortex neurons in very mild Alzheimer's disease. J Neurosci 60: 4491-4500

Gomez-Isla T, Hollister R, West H, Mui S, Growdon JH, Peterson RC, Parisi JE, Hyman BT (1997) Neuronal loss correlates with but exceeds neurofibrillary tangles in Alzheimer's disease. Ann Neurol 41: 17-24

Hung AY, Haass C, Nitsch RM, Qui WQ, Citron M, Wurtmann RJ, Growdon JH, Selkoe DJ (1993) Activation of protein kinase C inhibits cellular production of the amyloid β-protein. J Biol Chem 268: 22959-22962

Kovacs DM, Fausett HJ, Page KJ, Kim TW, Moir RD, Merriam DE, Hollister RD, Hallmark OG, Mancini R, Felsenstein KM, Hyman BT, Tanzi RE, Wasco W (1996) Alzheimer-associated presenilin 1 and 2: Neuronal expression in brain and localization to intracellular membranes in mammalian cells. Nature Med 2: 224-229

Liang P, Pardee AB (1992) Differential display of eukaryotic messenger RNA by means of the polymerase chain reaction. Science 257: 967-971

Nitsch RM, Slack BE, Wurtman RJ, Growdon JH (1992) Release of Alzheimer amyloid precursor derivatives stimulated by activation of muscarinic acetylcholine receptors. Science 258: 304-307

Nitsch RM, Farber SA, Growdon JH, Wurtman RJ (1993) Release of amyloid beta-protein precursor derivatives by electrical depolarization of rat hippocampal slices. Proc Natl Acad Sci USA 90: 5191-5193

Nitsch RM, Growdon JH (1994) Role of neurotransmission in the regulation of amyloid beta-protein precursor processing. Biochem Pharmacol 47: 1276-1284

Nitsch RM, Deng M, Growdon JH, Wurtman RJ (1996) Serotonin 5-HT2a and 5-HT2c receptors stimulate amyloid precursor protein ectodomain secretion. J Biol Chem 271: 4188-4194

Nitsch RM, Deng M, Wurtmann RJ, Growdon JH (1997) Metabotropic glutamate receptor subtype mGlu R1a stimulates the secretion of the amyloid β-protein precursor ectodomain. J Neurochem 69: 704-712

O'Brien TM, Yang GP, Sanders L, Lau LF (1990) Expression of cyr61, a growth factor-inducible immediate-early gene. Mol Cell Biol 10: 3569-3577

Palmer AM, Francis PT, Benton JS, Sims NR, Mann DM, Neary D, Snowdon JS, Bowen DM (1987) Presynaptic serotonergic dysfunction in patients with Alzheimer's disease. J Neurochem 48: 8-15

Procter AW, Palmer AM, Francis PT, Lowe SL, Neary D, Murphy E, Doshi R, Bowen DM (1988) Evidence of glutamatergic denervation and possible abnormal metabolism in Alzheimer's disease. J Neurochem 50: 790-802

Scheuner D, Eckman C, Jensen M, Song X, Citron M, Suzuki N, Bird TD, Hardy J, Hutton M, Kukull W, Larson E, Levy-Lahad E, Viitanen M, Peskind E, Poorkaj P, Schellenberg G, Tanzi RE, Wasco W, Lannfelt L, Selkoe D, Younkin S (1996) Secreted amyloid β-protein similar to that in the senile plaques of Alzheimer's disease is increased in vivo by the presenilin 1 and 2 and APP mutations linked to familial Alzheimer's disease. Nature Med 2: 864-870

Seeger M, Norstedt C, Petanceska S, Kovacs DM, Gouras GK, Hahne S, Fraser P, Levesque L, Czernik AJ, St George Hyslop P, Sisodia SS, Thinakaran G, Tanzi RE, Greengard P, Gandy S (1997) Evidence for phosphorylation and oligomeric assembly of presenilin 1. Proc Natl Acad Sci USA 94: 5090-5094

Sherrington R, Rogaev EI, Liang Y, Rogaeva EA, Levesque G, Ikeda M, Chi H, Lin C, Li G, Holman K, Tsuda T, Mar L, Foncin JF, Bruni AC, Montesi MP, Sorbi S, Rainero I, Pinessi L, Nee L, Chumakov I, Pollen D, Brookes A, Sanseau P, Polinsky RJ, Wasco W, Da Silva HAR, Haines JL, Pericak-Vance MA,

Tanzi RE, Roses AD, Fraser PE, Rommens JM, St George-Hyslop PH (1995) Cloning of a gene bearing missence mutations in early onset familial Alzheimer's disease. Nature 375: 754–760

Sims NR, Bowen DM, Allen SJ, Smith CCT, Neary D, Thomas DJ, Davison AN (1983) Presynaptic cholinergic dysfunction in patients with Alzheimer's disease. J Neurochem 40: 503–509

Sock E, Leger H, Kuhlbrodt K, Schreiber J, Enderich J, Richter-Landsberg C, Wegner M (1997) Expression of Krox proteins during differentiation of the O-2A progenitor cell line CG-4. J Neurochem 68: 1911–1919

Suggs SV, Katzowitz JL, Tsai-Morris C, Sukhatme VP (1990) cDNA sequence of the human cellular early growth response gene Egr-1. Nucleic Acids Res 18: 4283

Thinakaran G, Borchelt DR, Lee MK, Slunt HH, Spitzer L, Kim G, Ratovitsky T, Davenport F, Nordstedt C, Seeger M, Hardy J, Levey AI, Gandy SE, Jenkins NA, Copeland NG, Price DL, Sisodia SS (1996) Endoproteolysis of presenilin 1 and accumulation of processed derivatives in vivo. Neuron 17: 181–190

Walter J, Grünberg J, Capell A, Peshold B, Schindzielorz A, Citron M, Mendla K, St George Hyslop P, Multhaup G, Selkoe DJ, Haass C (1997) Proteolytic processing of the Alzheimer's disease-associated presenilin. 1 generates an *in vivo* substrate for protein kinase C. Proc Natl Acad Sci USA 94: 5349–5354

Neuronal Regulation of Presenilin-1 Processing

H. Hartmann, J. Busciglio and B. A. Yankner[*]

Summary

Presenilin 1 (PS1) is ubiquitously expressed but causes pathology only in the brain. The basis of this neuropathogenic specificity is unknown. In non-neuronal cells, PS1 is localized predominantly in the endoplasmic reticulum, where it co-localizes with the resident endoplasmic reticulum protein BiP. In cultured human cortical neurons, PS1 also appears in neuritic processes, where it does not co-localize with BiP. This neuron-specific localization of PS1 is accompanied by a neuron-specific pattern of PS1 processing. PS1 is constitutively cleaved to 30 kD N- and 20 kD C-terminal fragments in all non-neuronal cell types examined, as well as in undifferentiated neurons and PC12 cells. An alternative pathway of PS1 proteolytic processing, which gives rise to 36 kD N- and 14 kD C-terminal fragments, appears in the brain and in cultured differentiated neurons and PC12 cells. The alternative pathway of PS1 cleavage is present at low levels during fetal development and is markedly induced postnatally. This pattern of alternative PS1 cleavage is not detected at significant levels in astrocytes, non-neural cells or peripheral tissues. Inhibitors of serine, cysteine and lysosomal proteases do not affect the constitutive or alternative PS1 cleavages. However, the specific proteasome inhibitor lactacystin increases the level of full-length PS1 without affecting the generation of the major cleavage products. These results suggest that full-length PS1 may undergo two pathways of proteolysis: limit degradation by proteasome or discrete proteolytic cleavage to stable fragments by as yet unknown proteases. Thus, PS1 processing is a dynamic process that has undergone specific adaption in neurons. Neuron–specific processing of PS1 may play a role in the brain-specific pathological effects of PS1 mutations in familial Alzheimer's disease.

---

[*] Department of Neurology, Harvard Medical School and The Children's Hospital, Enders 260, 300 Longwood Avenue, Boston, MA 02115.

Introduction

Mutations in the presenilin family of genes is now recognized as a major genetic cause of early-onset Alzheimer's disease (AD; Schellenberg et al. 1995; Levy-Lahad et al. 1995; Alzheimer's Disease Collaborative Group 1995). More than 25 inherited mutations in presenilin 1 (PS1) have been identified in families that exhibit autosomal dominant inheritance of AD; two mutations have been identified in the highly homologous protein presenilin 2 (PS2). A presenilin homologue in *C. elegans, sel-12*, has been identified and has been genetically implicated in Notch/lin-12 signaling (Levitan and Greenwald 1995), although the role of PS1 in this signaling pathway is unclear. Since PS1 has been localized predominantly to the endoplasmic reticulum (ER; Kovacs et al. 1995), PS1 may function to regulate the processing and transport of receptors and other cell surface proteins during their transit through the secretory pathway. This possibility is supported by the observation that PS1 mutations can alter the processing of the amyloid precursor protein (APP), resulting in increased production of the amyloidogenic 42 amino acid form of amyloid β protein (Aβ; Scheuner et al. 1996; Duff et al. 1996; Borchelt et al. 1996b; Citron et al. 1997).

Presenilins and APP are ubiquitously expressed proteins, yet inherited mutations specifically affect the brain causing AD. In the case of APP, this is presumably the result of brain-specific processing of APP or its metabolites, resulting in parenchymal deposition of Aβ in the brain. Given this precedent, we explored the possibility that PS1 may also exhibit a brain-specific pattern of processing that could underlie the brain-specific pathological changes caused by PS1 mutations. PS1 is proteolytically processed to stable N- and C-terminal fragments in all cell types examined (Thinakaran et al. 1996). We have shown that this pattern of cleavage is altered in the adult brain and in differentiated neurons in culture, resulting in the production of alternative N- and C-terminal PS1 fragments (Hartmann et al. 1997). Here we investigate further the differential localization of PS1 in neuronal and non-neuronal cells and the developmental regulation of PS1 cleavage in the brain. Neuron-specific cleavage products of PS1 appear at low levels during fetal development but increase markedly in the adult brain. The generation of stable PS1 fragments by discrete proteolysis appears to be competing with a pathway of limit degradation mediated by proteasome. Thus, proteolysis of PS1 is precisely controlled and differentially regulated in the developing and adult brain, consistent with a functional role for PS1 in developing and mature neurons.

Results

To compare the localization of PS1 in neuronal and non-neuronal cells, we performed double-label immunofluorescence microscopy for PS1 and the ER marker BiP (Busciglio et al. 1997; Hartmann et al. 1997). Immunofluorescence microscopy of PS1-transfected COS cells using Ab231 directed against the PS1

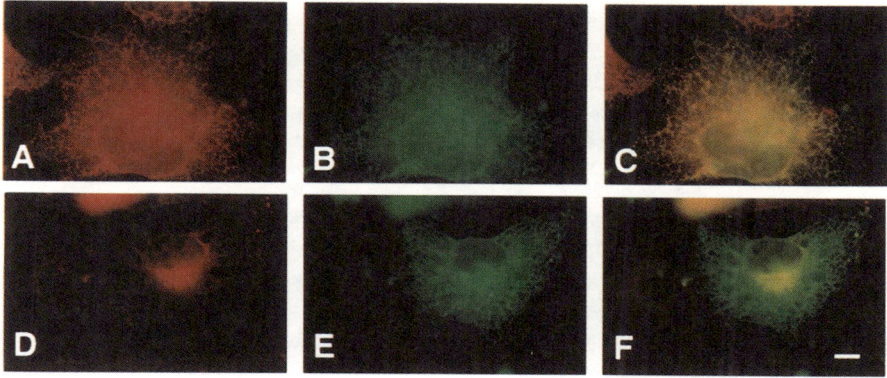

Fig. 1. Localization of transfected PS1 in non-neuronal cells. COS cells were examined 48 hrs after transfection with the PS1 cDNA by double-label immunofluorescence microscopy with Ab 231 to PS1 (**A**, green) and anti-BiP, a marker of the rough ER (**B**, red). Overlap of the two labels is yellow (**C**). Transfected COS cells were also double-labeled with Ab 231 (**D**) and antibody JE4, a marker of the Golgi complex (**E**); a minor component of PS1 co-localizes with the Golgi marker in transfected cells (**F**, yellow)

N-terminus demonstrated a reticular and perinuclear distribution of PS1 that co-localized with BiP (Fig. 1A–C). A minor component of PS1 also co-localized with a marker for the Golgi complex (Fig. 1D–F). The specificity of staining was confirmed by preabsorption of Ab231 with the antigenic PS1 peptide, which abolished immunoreactivity (data not shown). These results confirm the previously reported ER and Golgi distribution of transfected PS1 (Kovacs et al. 1996). We then examined the subcellular distribution of endogenous PS1 in primary human cortical cultures. PS1 appeared in both the cell body and distal neuritic processes of human cortical neurons (Fig. 2). In contrast, immunoreactivity for the ER marker BiP was restricted to the cell soma and some proximal neurites. PS1 and

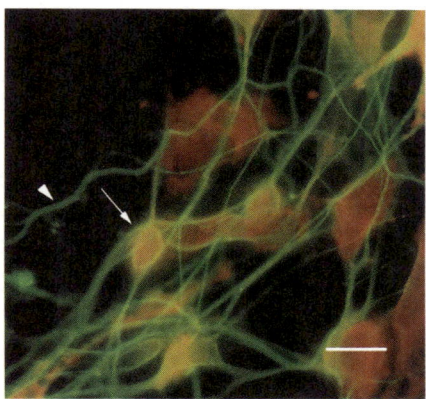

Fig. 2. Localization of endogenous PS1 in cultured human cortical neurons. Primary cultures of human fetal cortex were maintained in vitro for seven days an double-labeled with Ab 231 to PS1 (green) and anti-BiP (red). PS1 in neuronal cell bodies (arrow) co-localizes with the ER marker BiP (overlap of PS1 and BiP is yellow). PS1 in neurites does not overlap with BiP (arrowhead). PS1 staining in neuronal cell bodies and neurites was abolished by preabsorption of Ab 231 with the cognate PS1 peptide (not shown). Astrocytes are the large cells that stain red for BiP; astrocytes do not show significant PS1 staining. Scale bar: 10 µM

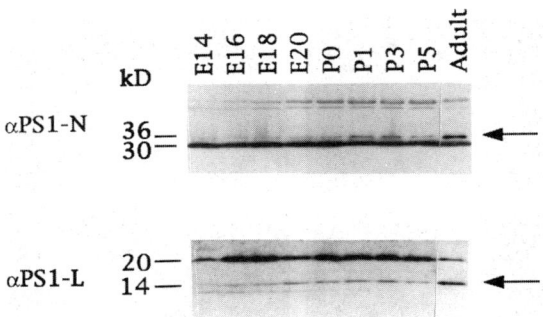

Fig. 3. Regulation of PS1 proteolytic cleavage during rat brain development. Shown is Western blot analysis of rat brain homogenates with antibodies to the PS1 N-terminus (αPS1-N) and the large C-terminal loop (αPS1-L). Embryonic day (E) 15, 16, 18, 20; postnatal day (P) 0, 1, 3, 5; 3-month-old rat cortex (adult). Note the induction of 36 kD N-terminal and 14 kD C-terminal fragments (arrows) at late stages of development and predominantly in the adult

BiP co-localized in the cell soma but not in the distal neurites (Fig. 2, overlap indicated by yellow). Thus, neuronal PS1 co-localizes with a marker of the rough ER in the cell body but not in the neurites.

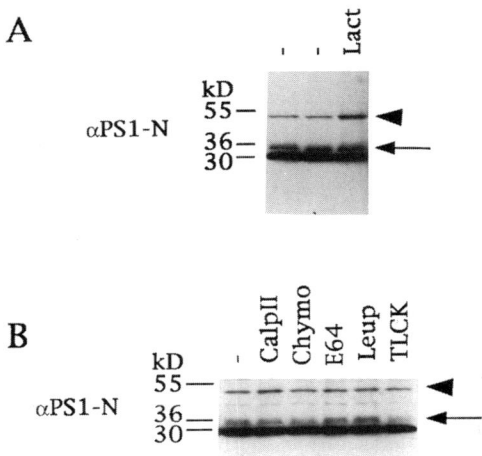

Fig. 4. Proteasome degrades full-length PS1. PC12 cells were differentiated by treatment with nerve growth factor (50 ng/ml) for three days. Cell lysates were analyzed by Western blotting with αPS1-N. (A) Differentiated PC12 cells treated with the proteasome inhibitor lactacystin (Lact; 10 μM for 14 hrs) show increased full-length PS1 (arrowhead) relative to untreated cultures (–). Levels of N-terminal fragments (arrow) are unaffected. (B) Differentiated PC12 cells were untreated (–) or incubated for 14 hrs with calpain inhibitor II (CalpII; 25 μM), chymostatin (Chymo; 60 μg/ml), E64 (E64; 10 μg/ml), leupeptin (Leup; 500 μg/ml) and TLCK (TLCK; 37 μg/ml). Levels of full-length PS1 (arrowhead) and N-terminal fragments (arrow) were not significantly affected

PS1 undergoes constitutive proteolytic cleavage to 30 kD N- and 20 kD C-terminal fragments in non-neuronal cells (Thinakaran et al. 1996). However, in the adult rat brain and in cultured rat hippocampal neurons, PS1 undergoes an additional proteolytic cleavage that generates 36 kD N- and 14 kD C-terminal fragments (Hartmann et al. 1997). To examine the regulation of PS1 cleavage in the rat brain during fetal development, Western blot analysis was performed from embryonic day 12 (E12) to postnatal day 5 (P5). The constitutive 30 kD N-terminal and 20 kD C-terminal fragments were the predominant PS1 species throughout embryonic development and in early postnatal life in the rat brain (Fig. 3). The alternative 36 kD N- and 14 kD C-terminal fragments were present at low levels throughout development and gradually increased from E20 to P5. A much greater increase in the alternative PS1 fragments appeared after P5 in the adult rat brain (Fig. 3). These results suggest that alternative proteolytic cleavage of PS1 occurs predominantly in the postnatal rat brain.

To investigate the proteolytic systems involved in the processing of PS1, we examined the effects of a variety of proteases on PS1 processing in PC12 cells. Undifferentiated rat PC12 cells exhibit the constitutive cleavage pattern characteristic of non-neuronal cells. When PC12 cells are differentiated with nerve growth factor, the alternative PS1 fragments appear (Hartmann et al. 1997). Incubation of differentiated PC12 cells with the serine protease inhibitor TLCK, the cysteine protease inhibitors leupeptin and E64, the chymotrypsin inhibitor chymostatin or a calpain inhibitor did not significantly affect the generation of either the constitutive or alternative PS1 fragments (Fig. 4B). We then examined the effect of lactacystin, a highly specific inhibitor of the proteasome (Fenteany et al. 1995). Treatment with lactacystin significantly increased the level of full-length PS1, but did not affect the levels of the N-terminal fragments (Fig. 4A). In addition, lactacystin did not significantly affect the levels of PS1 C-terminal fragments (data not shown). These results suggest that proteasome mediates the limit degradation of full-length PS1, but does not mediate the constitutive or alternative proteolytic cleavages.

Discussion

These experiments indicate that PS1 is localized to both the cell body and neuritic processes of cultured human cortical neurons. It is noteworthy that PS1 in neurites does not co-localize with the ER marker BiP, in contrast to the co-localization of PS1 with BiP in non-neuronal cells. This difference may reflect the presence of PS1 in a compartment other than the ER in neurons, or in a compartment of the smooth ER that does not contain BiP. A recent ultrastructural study of PS1 localization in the brain showed that PS1 is localized to both tubular and vesicular compartments in neurons and suggested that PS1 may be part of the intermediate compartment between the ER and Golgi (Lah et al. 1997). The intermediate compartment has been implicated in the regulation of protein transport and may also be involved in protein degradation or processing. As such, this compartment could potentially be the site of action of PS1 mutations that alter APP processing, giving rise to increased levels of Aβ42 (Scheuner et al. 1996). However, the absence of rough ER and Golgi complex in neuritic processes raises the possibility that neuronal PS1 may be involved in functions other than protein processing.

The predominant PS1 species appear to be N- and C-terminal fragments in a variety of cell types (Thinakaran et al. 1996). The full-length PS1 protein is detected at only low levels in non-transfected cells. Thus, the regulation of PS1 processing is likely to impact on its biological activity. We have demonstrated that the cleavage of PS1 is regulated by neuronal differentiation (Hartmann et al. 1997). Differentiation of neuronal cells in culture is accompanied by the generation of longer N-terminal and shorter C-terminal PS1 fragments, in addition to the constitutive fragments. This pattern is recapitulated *in vivo*; alternative PS1 fragments appear at low levels in the fetal brain and are significantly induced in the adult brain. In both rat and human, alternative cleavage of PS1 occurs predominantly in the brain; much lower levels of these fragments are detected in peripheral tissues. Thus, generation of the alternative PS1 fragments may be related to a neuron-specific function of PS1. Altered cleavage of PS1 may therefore be a mechanism for regulating its biological activity in a cell type-specific context.

The single major alternative N-terminal fragment detected in the rat brain is in contrast to the ladder of alternative N-terminal fragments in the human brain (Hartmann et al. 1997). Furthermore, a major alternative C-terminal fragment is detected in the rat brain, but alternative C-terminal fragments are not detected in the human brain. It is possible that the heterogeneous cleavage of PS1 in the human brain results in rapid degradation of the truncated C-terminal fragments. The alternative cleavage pathway may therefore serve to regulate the stability and biological activity of PS1 fragments.

The proteases responsible for the constitutive and alternative cleavages of PS1 remain to be identified. A variety of serine and cysteine protease inhibitors had no effect on PS1 processing. However, the highly specific proteasome inhibitor lactacystin increased the steady state level of full-length PS1 in PC12 cells, without altering the levels of the N- or C-terminal fragments. Thus, the proteasome appears to mediate limit degradation of full-length PS1 but does not significantly affect the major cleavage pathways. Interestingly, the increased levels of full-length PS1 generated in the presence of the proteasome inhibitor did not give rise to increased levels of the major PS1 fragments, suggesting that the fragment-generating pathway is tightly regulated. It remains to be determined whether the PS1 mutations act at the level of PS1 processing. A consistent effect of pathogenic PS1 mutations on PS1 processing has not been established. Various PS1 mutations have been reported to either inhibit PS1 cleavage in vitro (Mercken et al. 1996; Thinakaran et al. 1996) or increase PS1 cleavage in transgenic mice (Borchelt et al. 1996b). It will be of interest to determine whether PS1 mutations affect the neuron-specific alternative cleavage pathway. A selective effect of PS1 mutations on neuron-specific processing could potentially explain the brain specificity of pathology associated with PS1 mutations.

References

Alzheimer's Disease Collaborative Group (1995) The structure of the presenilin 1 (S182) gene and identification of six novel mutations in early onset AD families. Nature Genet 11: 219–222

Borchelt D, Lee M, Thinakaran G, Slunt HH, Ratovitski T, Kim G, Levey A, Gandy S, Jenkins NA, Copeland NG, Price DL, Sisodia SS (1996) Endoproteolysis and saturable accumulation of human presenilin 1 derivatives in transgenic mice. Abstr Soc Neurosci 11: 728

Borchelt DR, Thinakaran G, Eckman CB, Lee MK, Davenport F, Ratovitsky T, Prada C-M, Kim G, Seekins S, Yager D, Slunt HH, Wang R, Seeger M, Levey AI, Gandy SE, Copeland NG, Jenkins NA, Price DL, Younkin SG, Sisodia SS (1996) Familial Alzheimer's disease-linked presenilin 1 variants elevate Aβ1–42/1–40 ratio in vitro and in vivo. Neuron 17: 1005–1013

Busciglio J, Hartmann H, Lorenzo A, Wong C, Baumann K-H, Sommer B, Staufenbiel M, Yankner BA (1997) Neuronal localization of presenilin-1 and association with amyloid plaques and neurofibrillary tangles in Alzheimer's disease. J Neurosci 17: 5101–5107

Citron M, Westaway D, Xia W, Carlson G, Diehl T, Levesque G, Johnson-Wood K, Lee M, Seubert P, Davis A, Kholodenko D, Motter R, Sherrington R, Perry B, Yao H, Strome R, Lieberburg I, Rommens J, Kim S, Schenk D, Fraser P, St George Hyslop P, Selkoe DJ (1997) Mutant presenilins of Alzheimer's disease increase production of 42-residue amyloid β-protein in both transfected cells and transgenic mice. Nature Med 3: 67–72

Fenteany G, Standaert RF, Lane WS, Choi S, Corey EJ, Schreiber SL (1995) Inhibition of proteasome activities and subunit-specific amino-terminal threonine modification by lactacystin. Science 268: 726–731

Hartmann H, Busciglio J, Baumann K-H, Staufenbiel M, Yankner BA (1997) Developmental regulation of presenilin-1 processing in the brain suggests a role in neuronal differentiation. J Biol Chem 272: 14505–14508

Kovacs DM, Fausett HJ, Page KJ, Kim T-W, Moir RD, Merriam DE, Hollister RD, Hallmark OG, Mancini R, Felsenstein KM, Hyman BT, Tanzi RE, Wasco W (1996) Alzheimer-associated presenilins 1 and 2: neuronal expression in brain and localization to intracellular membranes in mammalian cells. Nature Med 2: 224–229

Lah JJ, Heilman CJ, Nash NR, Rees HD, Yi H, Counts SE, Levey AI (1997) Light and electron microscopic localization of presenilin-1 in primate brain. J Neurosci 17: 1971–1980

Levitan D, Greenwald I (1995) Facilitation of *lin 12* mediated signalling by *sel-12*, a *Caenorhabditis elegans* S182 Alzheimer's disease gene. Nature 377: 351–354

Levy-Lahad E, Wasco W, Poorkaj P, Romano DM, Oshima J, Pettingell WH, Yu C-E, Jondro PD, Schmidt SD, Wang K, Crowley AC, Fu Y-H, Guenette SY, Galas D, Nemens E, Wijsman EM, Bird TD, Schellenberg GD, Tanzi RE (1995) Candidate gene for the chromosome 1 familial Alzheimer's disease locus. Science 269: 973–977

Mercken M, Takahashi H, Honda T, Sato K, Murayama M, Nakazato Y, Noguchi K, Imahori K, Takashima A (1996) Characterization of human presenilin 1 using N-terminal specific monoclonal antibodies: Evidence that Alzheimer mutations affect proteolytic processing. FEBS Lett 389: 297–303

Scheuner D, Eckman M, Jensen X, Song M, Citron N, Suzuki TD, Bird J, Hardy M, Hutton W, Kukull E, Larson E, Levy-Lahad M, Viitanen E, Peskind P, Poorkaj G, Schellenberg R, Tanzi W, Wasco L, Lannfelt L, Selkoe D, Younkin S (1996) Secreted amyloid β-protein similar to that in the senile plaques of Alzheimer's disease is increased in vivo by the presenilin 1 and 2 and APP mutations linked to familial Alzheimer's disease. Nature 2: 864–870

Sherrington R, Rogaev EI, Liang Y, Rogaeva EA, Levesque G, Ikeda M, Chi H, Li G, Holman K, Tsuda T, Mar L, Foncin J-F, Bruni AC, Montesi MP, Sorbi S, Rainero I, Pinessi L, Nee L, Chumakov I, Pollen D, Brookes A, Sanseau P, Polinsky RJ, Wasco W, Da Silva HA, Haines JL, Pericak-Vance MA, Tanzi RE, Roses AD, Fraser PE, Rommens JM, St George Hyslop PH (1995) Cloning of a gene bearing missense mutations in early-onset familial Alzheimer's disease. Nature 375: 754–760

Thinakaran G, Borchelt DR, Lee MK, Ratovitsky T, Davenport F, Nordstedt C, Seeger M, Hardy J, Levey A, Gandy SE, Jenkins NA, Copeland NG, Price DL, Sisodia SS (1996) Endoproteolysis of presenilin 1 and accumulation of processed derivatives in vivo. Neuron 17: 181–190

Transgenic Approaches to the Study of Alzheimer's Disease

K. Duff[*]

Summary

Different genes known to be involved in Alzheimer's disease have been introduced into transgenic mice in an attempt to model the disease. Transgenic mice overexpressing mutant presenilin 1 (PS1) transgenes show elevated levels of the highly amyloidogenic fragment, Aβ(42)43, compared to mice overexpressing wild type PS1. Mutations in both PS1 and APP have therefore been shown to have a similar effect on Aβ(42)43 levels suggesting a common link to AD pathogenesis. Although mutant PS1 and APP transgenic mice do not show all the features of the human disease, it is now possible to extend and enhance the AD phenotype by crossing transgenic mice together.

APP Transgenic Models

The last two years have seen the resurgence in popularity of the transgenic mouse as a model system in which to study Alzheimer's disease. In 1995, Athena Neuroscience published their groundbreaking transgenic APP mouse model, PDAPP (Games et al. 1995). This mouse consisted of an APP minigene complete with introns that allowed splicing of the cDNA to generate the three most abundant isoforms of APP in the brain. The construct was under the control of the PDGF promoter and expressed the V717F mutation. The PDAPP mouse demonstrated beta amyloid (Aβ) containing deposits in the cortex and hippocampus that stained with thioflavin-S and congo red and showed reactive gliosis surrounding the deposits, as well as other markers of cellular disturbance. Later work by the Athena group carefully examined APP levels, the correlation of Aβ levels with deposit formation and the spread of the pathology to different regions of the brain with age in heterozygous and homozygous PDAPP mice (Johnson-Wood et al. 1997). In 1996, Karen Hsiao and colleagues published a report of a second transgenic mouse with Aβ deposits where the formation of deposits correlated with elevated Aβ levels and behavioral impairments (Hsaio et al. 1996). This mouse used the hamster PrP promoter to drive the expression of an NL 670/671 APP695 cDNA.

---

[*] Mayo Clinic Jacksonville, 4500 San Pablo Rd., Jacksonville, FL 32224, USA

These papers showed that transgenic mice overexpressing APP could be induced to develop amyloid deposits when the level of Aβ was high enough. It appears, therefore, that Aβ levels in the brain must reach a theoretical "threshold level" of Aβ for the deposits to form and this level appears to be above 50 pm/g total Aβ. This would explain the lack of deposits in previous transgenic mice where the levels of Aβ presumably fell below the threshold.

PS1 Transgenic Models

One month after the publication of the Tg2576 mouse, we reported the creation of transgenic mice overexpressing the human PS1 cDNA (Duff et al. 1996). These mice used the PDGF promoter to drive the expression of the PS1 cDNA. Three different mouse lines were created: wild type (WT), mutant M146L and mutant M146V. An intron from human genomic PS1 was inserted into the equivalent locus between exons 4 and 5 in the cDNA construct and this clearly had a great effect on the levels of transcription from the PDGF promoter. The PS1 protein was translated and processed correctly but the levels of protein in each mouse line were different, although this did not reflect the transgene copy number. The pattern of expression of the transgene was assessed by in situ hybridization and was found to be mainly neuronal as expected for the PDGF promoter (unpublished data).

An earlier study by Younkin and colleagues (Scheuner et al. 1996) had demonstrated that Aβ42(43) levels were elevated in the fibroblasts of patients with PS1 mutations compared to their unaffected relatives. We therefore decided to analyze the levels of endogenous mouse Aβ1–40 and 1–42(43) in the WT and mutant lines of mice. Our data confirmed the fibroblast study in that the levels of Aβ42(43) were elevated in mutant but not WT overexpressing mice, but the levels of Aβ1–40 were not (Duff et al. 1996). This result was subsequently confirmed in other transgenic mice and transfected cell lines (Borchelt et al. 1996; Thinakaran et al. 1996; Citron et al. 1997).

Other than the effect on Aβ generation, the PS transgenic mice have so far failed to show any overt AD phenotype. There is no obvious plaque or tangle pathology in aged mice, and they do not show any behavioral impairments (unpublished data). A battery of tests is, however, currently underway, both on the mice and on primary neuronal cultures derived from them, to identify any cellular defect that could explain the role of PS in normal cellular function and in AD. These tests include looking for markers of oxidative damage (such as MDA or SOD overproduction), markers of apoptotic cell death, alterations in brain electrophysiology coupled with calcium imaging, and investigating the processing of PS1 itself. It is possible though that the only contribution of the presenilin mutations is to divert APP processing down a more amyloidogenic pathway, either due to a direct interaction between the two proteins or to some role of PS on APP trafficking.

Modulating AD Phenotype in Transgenic Mice

The intriguing observation that PS1 mutations affect APP processing in such a way that more of the amyloidogenic fragment Aβ1-42(43) is generated is highly suggestive of the central role of this molecule in AD pathology. One way to test this theory is to cross an APP transgenic mouse that develops deposits with a PS1 mutant mouse and study the effects of the two transgenes on pathogenesis. It is predicted that the marginal initial increase in Aβ42(43) levels in PS/APP mice relative to APP littermates will accelerate the rate at which Aβ deposits develop if deposition is related to overall Aβ levels, as suggested by Athena's studies on homozygous vs. heterozygous PDAPP mice. Two recent publications have shown that this is in fact the case (Borchelt et al. 1997, Holcomb et al. 1998) with increases in Aβ(42)43 as low as 40% accelerating pathology development from 12 months (in the singly transgenic APP animal) to 12 weeks in the doubly transgenic PS/APP animal (Holcomb et al. 1998 and Duff, unpublished data). It will be interesting to correlate Aβ elevation with deposit formation and to assess the "accelerated" mice for other AD-related markers such as neuronal loss or tangle pathology that may form due to the increased period of exposure to toxic Aβ.

Crossing different transgenic mice together may help to modulate the phenotype in several ways. For example, abnormal tau pathology appears to be a significant feature of AD but is is unlikely to be causative, at least not in AD caused by APP mutations. The association of abnormal tau pathology with several neurodegenerative diseases (reviewed in Delacourte and Buée 1997) suggests that the development of paired helical filaments (PHF) and tangles is detrimental to neuronal survival, and the lack of widespread neuronal loss in APP overexpressing mice might at least in part be related to a lack of robust tau pathology. It does appear though, that some abnormal tau pathology has been achieved in aged, APP overexpressing mice (B. Sommer, personal communication), suggesting that these structures may be a consequence of APP/Aβ damage. If human tau overexpressing mice can be crossed with APP overexpressing mice, the combination of high levels of human tau in the disturbed cellular environment created by APP/Aβ overexpression might stimulate the formation of robust tangles and we might begin to see neuronal degeneration. Alternatively, if oxidative stress is a factor in AD pathogenesis, crossing APP mice with mice overexpressing protective agents (such as SOD) may reduce the impact of the deposits on cellular dysfunction or may retard the development of deposits themselves.

The issue of cognitive impairment in deposit forming mice is also central to our understanding of the role of deposits in dementia associated with Alzheimer's disease. As both the APP transgenic models generated so far fail to show widespread neuronal loss, it is intriguing that cognitive impairment has been reported in one of them (Hsiao et al. 1996). It will be interesting to examine whether the impairment precedes deposit formation, as the sequence of events in AD pathogenesis will have significant implications in drug design. It is possible that soluble (or more likely, partially fibrillar) Aβ is the cytotoxic entity that is responsible for cognitive impairment and the deposition of amyloid into plaques might be a pathogenically unrelated consequence of high concentrations of Aβ or, indeed, might be a protective cellular mechanism that clears the brain of toxic fibrils. Alternatively, APP overexpression may be the culprit, or some early Aβ related intracellular event.

Overview

Although it would be premature to predict that the specific elevation of Aβ1-42(43) relative to other forms of Aβ peptide is the pathogenic mechanism in Alzheimer's disease, it does seem remarkable that mutations in three genes (PS1, PS2 and APP) all affect the same pathway. The observation that elevation above a certain threshold is also linked to amyloid deposition in the APP mice, that Aβ1-42(43) maybe at the root of these features of the disease. The exact sequence of events is still unclear but the transgenic models generated so far have already shown their utility in dissecting this complex part of the pathology, and the potential for crossing mice reinforces the validity of the transgenic approach to studying human neurodegenerative diseases.

References

Borchelt DR, Ratovitski T, van Lare J, Lee MK, Gonzales V, Jenkins NA, Copeland NG, Price DL, Sisodia SS. Accelerated amyloid deposition in the brains of transgenic mice coexpressing mutant presenilin 1 and amyloid precursor proteins. Neuron 1997 Oct; 19(4): 939–945

Borchelt DR, Thinakaran G, Eckman CB, Lee MK, Davenport F, Ratovisky T, Prada CM, Kim G, Seekins S, Yager D, Slunt HH, Wang R, Seeger M, Levey M, Levey AI, Gandy SE, Copeland NG, Jenkins NA, Price DL, Younkin SG, Sisodia SS (1996) Familial Alzheimer's disease-linked presenilin 1 variants elevate Aβ1-42/1-40 ratio in vitro and in vivo. Neuron 17: 1005–1013

Citron M, Westaway D, Xia W, Carlson G, Diehl T, Levesque G, Johnson-Wood K, Lee M, Seubert P, Davis A, Kholodenko D, Motter R, Sherrington R, Perry B, Yao H, Stome R, Lieberburg I, Rommens J, Kim S, Schenk D, Fraser P, St George Hyslop P, Selkoe DJ (1997) Mutant presenilins of Alzheimer's disease increase production of 42-residue amyloid β-protein in both transfected cells and transgenic mice. Nat Med 3: 67–68

Delacourte A, Buée L (1997) Normal and pathological tau proteins as factors for microtubule assembly. Int J Cytol 171: 167–223

Duff K, Eckman C, Zehr C, Yu X, Prada CM, Perez-Tur J, Hutton M, Buee L, Harigaya Y, Yager D, Morgan D, Gordon MN, Holcomb L, Refolo L, Zenk B, Hardy J, Younkin S (1996) Increased amyloid-β42(43) in brains of mice expressing mutant presenilin 1. Nature 383: 710–713

Games D, Adams D, Alessandrini R, Barbour R, Berthelette P, Blackwell C, Carr T, Clemena J, Donaldson T, Gillespie F, Guido T, Hagopian S, Johnson-Wood K, Khan K, Lee M, Leibowitz P, Leiberberg I, Little S, Masilah E, McConlogue L, Montoya-Zavala M, Mucke L, Paganini L, Penniman E, Power M, Schenk D, Seubert P, Snyder B, Soriano F, Tan H, Vitale J, Wadsworth S, Wolozin B, Zhao J (1995) Alzheimer-type neuropathology in transgenic mice overexpressing V717F β-amyloid precursor protein. Nature 373: 523–527

Holcomb L, Gordon M, McGowan E, Yu X, Benkovic S, Jantzen P, Wright K, Saad I, Mueller R, Morgan D, Sanders S, Zehr C, O'Campo K, Hardy J, Prada C, Eckman C, Younkin S, Hsiao K, Duff K (1998) Accelerated Alzheimer-type phenotype in transgenic mice carrying both mutant amyloid precursor protein and presenilin 1 transgenes. Nat. Med, Vol. 4 (No 1); p. 97;

Hsiao K, Chapman P, Nilsen S, Eckman C, Harigaya Y, Younkin S, Yang F, Cole G (1996) Correlative memory deficits, Aβ elevation and amyloid plaques in transgenic mice. Science 274: 99–102

Johnson-Wood K, Lee M, Motter R, Hu K, Gordon G, Barbour R, Khan K, Gordon M, Tan H, Games D, Lieberberg I, Schenk D, Seubert P, McConologue L (1997) Amyloid precursor protein processing and Aβ42 deposition in a transgenic mouse model of Alzheimer's disease. Proc Natl Acad Sci USA 94: 1550–1555

Scheuner D, Eckman C, Jensen M, Song X, Citron M, Suzuki N, Bird TD, Hardy J, Hutton M, Kukull W, Larson E, Levey-Lahad E, Viitanen M, Peskind E, Poorkaj P, Schellenberg G, Tanzi R, Wasco W, Lannfelt L, Selkoe D, Younkin S (1996) Secreted amyloid β-protein similar to that in the senile plaques of Alzheimer's disease is increased in vivo by the presenilin 1 and 2 and APP mutations linked to familial Alzheimer's disease. Nat Med 2: 864–870

Sturchler-Pierrat C, Abramowski D, Duke M, Wiederhold KH, Mistl C, Rothacher S, Ledermann B, Burki K, Frey P, Paganetti PA, Waridel C, Calhoun ME, Jucker M, Probst A, Staufenbiel M, Sommer B (1997) Two amyloid precursor protein transgenic mouse models with Alzheimer disease-like pathology. Proc Natl Acad Sci USA Nov 25; 94(24): 13287–13292.

Thinakaran G, Borchelt DR, Lee MK, Slunt HH, Spitzer L, Kim G, Ratovitsky T, Davenport F, Nordstedt C, Seeger M, Hardy J, Levey AI, Gandy SE, Jenkins NA, Copeland NG, Price DL, Sisodia SS (1996) Endoproteolysis of presenilin 1 and accumulation of processed derivatives in vivo. Neuron 17: 181–190

Subject Index

acetylcholine 79–85
AMPA glutamate receptor 37
β-amyloid 7, 12, 15, 16, 19, 20, 27–36, 43, 57, 62, 66, 67, 71–78, 80, 93, 94
β-amyloid precursor protein 1, 7, 11, 12, 16, 19, 20, 27, 31, 36, 38, 39, 44, 49–70, 79, 81, 82, 86, 90, 93–95
β-amyloid precursor protein processing 11, 20, 22, 27–33, 35, 44, 56, 62, 90, 94
amyloid cascade hypothesis 11
amyloid deposition 11, 15, 19, 20, 27–29, 31, 32, 36, 86, 93, 94
amyloid metabolism 27–33, 35, 49–58
amyloid toxicity 31
animal model 6, 7
α_1-anti-chymotrypsin 4
apolipoprotein E 1, 4, 11, 16–17, 19, 20, 29, 31, 59
apoptosis 23, 61, 62, 94

behavior 6, 7, 93, 95, 96
biology of the human presenilins 4–6, 20–23, 35–58, 60–62, 71–78, 85–91
BiP 39, 85–89

Caenorhabditis elegans 2, 12, 36, 37, 40, 41, 43, 60, 61, 66, 71–73, 86
casein kinase 74, 75
cerebrospinal fluid 27, 30
chromosome 1 1, 2, 19, 36
chromosome 12 4
chromosome 14 2, 36
chromosome 19 1, 19, 59
chromosome 21 1, 19, 36, 59
co-immunoprecipitation 38, 53, 54, 63, 64, 66, 67

detection of presenilin proteins in brain tissue 51, 52
Dll1 35, 42, 43
dominance 11–17

dominant negative 15
Down syndrome 19
Drosophila melanogaster 2

early growth response gene 79, 81, 82
early onset form of Alzheimer's disease 1–3, 11, 16, 27–33, 36, 59, 86
electrophysiology 7
embryogenesis 6, 35, 41–43, 73, 89
endoplasmic reticulum 5, 12, 20–22, 39, 53, 55, 56, 60, 66–68, 75, 76, 85–87, 89, 90
endoproteolysis of presenilin 6, 19–25, 35, 37–41, 43, 49–52, 55, 61, 67, 72, 85, 86, 89, 90
evolution 2
extracellular amyloid β-protein 27–33

familial Alzheimer's disease 1–10, 19–33, 35–37, 40, 41, 43, 49, 51–53, 59, 60, 62, 66, 67, 71–73, 81, 85
free radical 62
function of presenilin (cf. biology of presenilin)

G protein 80
gain of function 12–15, 35
gene expression 79, 81, 82
genetics 1–25, 27–30, 59, 86
Gig1 79, 82
glutamate 80
Golgi apparatus 5, 12, 20–22, 53, 55, 56, 60, 66–68, 75, 76, 87, 89, 90

intracellular amyloid β production 53–56

Jagged 42, 43

kainate binding protein 37

long-term depression (LTD) 7
long-term potentiation (LTP) 7
loss of function 3, 12-15, 35

metabolism of presenilin (cf. biology of presenilin)
muscarinic receptor 79-84
Δ_9 mutation 12, 15, 16, 36, 37, 41, 61
mutations in the presenilin genes 1-4, 12-17, 27-30, 49, 54, 60, 86

neuronal regulation of presenilin processing 85-91
neurotransmitter 79, 80
NMDA receptor 37
notch 12, 35, 36, 41-43, 71-74, 86

overproduction of β-amyloid42(43) 7, 11, 12, 16, 19, 20, 23, 27-33, 36, 43, 44, 52, 53, 55, 56, 61, 62, 66, 75, 76, 81, 86, 90, 94-96

pathogenesis of Alzheimer's disease 11-18, 22, 27, 43, 44, 62, 82, 85, 95
phosphorylation of presenilin 50, 74, 75, 79-84
presenilin 2 - APP interactions 59-76

presenilin knock out mice 35, 42, 43
presenilin transgenic mice 6, 7, 32, 93-96
protein kinase A 75
protein kinase C 75, 79-81
protein trafficking 7, 12, 20, 41, 44, 62, 66, 94

secretase 11, 36, 55, 62, 76, 79-81
sel 12 3, 12, 15, 16, 20, 36-38, 40, 41, 43, 61, 62, 71-74, 86
serotonin 80
somitic cell lineage 35, 42, 43
spe4 12, 20
sporadic Alzheimer's disease 30, 31, 55
structure of presenilins 2, 12-14, 20, 37, 41, 60, 68
swedish mutation 27-31, 52, 54

therapy 44, 79
transcriptional activation 79-84
transfected cell 5, 6, 12, 15, 21, 32, 50-55, 86, 87, 90, 94
transferrin receptor 56
transgenic mice 6, 7, 15, 22, 32, 36-39, 41, 43, 44, 53, 72, 90, 93-96

VLDL receptor 4
Volga German families 2, 22, 23, 30, 62, 67

Printing: Saladruck, Berlin
Binding: Buchbinderei Lüderitz & Bauer, Berlin